From Projects to Programs

A Project Manager's Journey

T0293722

Best Practices and Advances in Program Management Series

Series Editor
Ginger Levin

PUBLISHED TITLES

From Projects to Programs: A Project Manager's Journey
Samir Penkar

Sustainable Program Management
Gregory T. Haugan

Leading Virtual Project Teams: Adapting Leadership Theories and Communications Techniques to 21st Century Organizations
Margaret R. Lee

Applying Guiding Principles of Effective Program Delivery
Kerry R. Wills

Construction Program Management
Joseph Delaney

Implementing Program Management: Templates and Forms Aligned with the Standard for Program Management, Third Edition (2013) *and Other Best Practices*
Ginger Levin and Allen R. Green

Program Management: A Life Cycle Approach
Ginger Levin

FORTHCOMING TITLES

Successful Program Management: Complexity Theory, Communication, and Leadership
Wanda Curlee and Robert Lee Gordon

Program Management Leadership: Creating Successful Team Dynamics
Mark C. Bojeun

The Essential Program Management Office
Gary Hamilton

From Projects to Programs

A Project Manager's Journey

Samir Penkar, PMP, CSM

CRC Press
Taylor & Francis Group
Boca Raton   London   New York

CRC Press is an imprint of the
Taylor & Francis Group, an **informa** business

CRC Press
Taylor & Francis Group
6000 Broken Sound Parkway NW, Suite 300
Boca Raton, FL 33487-2742

© 2014 by Taylor & Francis Group, LLC
CRC Press is an imprint of Taylor & Francis Group, an Informa business

No claim to original U.S. Government works

Printed on acid-free paper
Version Date: 20130507

International Standard Book Number-13: 978-1-4665-9181-3 (Paperback)

Library of Congress Cataloging-in-Publication Data

Penkar, Samir.
  From projects to programs : a project manager's journey / Samir Penkar, PgMP.
    pages cm
  Includes bibliographical references and index.
  ISBN 978-1-4665-9181-3
  1. Project management. I. Title.

HD69.P75P457 2013
658.4'04--dc23   2013017420

Visit the Taylor & Francis Web site at
http://www.taylorandfrancis.com

and the CRC Press Web site at
http://www.crcpress.com

Contents

Preface

This is a story about a project manager growing into a program manager's shoes. It is a chronicle of a program manager's first program—her growth, struggles, and wins as she navigates this complex area. It is in the form of a story, told from the program manager's point of view. I did this because I have found that people relate better to stories; they can connect to the concepts in a better way and then relate it back to their own situations.

Whatever your goal, whether it be moving up the project management career ladder, understanding program management, or just looking for a refreshing business book, I hope *From Project to Programs: A Project Manager's Journey* can help you with *your* journey. Written in a conversational tone, you will gain insights into the mind of a program manager, a peek into her personal life, and how work and life are so intertwined. Throughout this story, you will see yourself somewhere in these pages. Pause, stop, and ponder on the reflection questions at the end of each chapter. Think what you would have done in this situation. This is not an attempt to be a comprehensive guide to program management, but rather an attempt to distill some of the core areas of program management.

You will see my love for running, visual thinking, and music in these pages. This book is a journey into the future, and a journey of progress. I hope and trust you enjoy it as much as I have enjoyed writing it.

Acknowledgments

To Aie, Papa, Amita, and my little Manas

Writing a book is such a joy and challenge at the same time. You see your creation come to life and realize it takes more than you to create it. It takes the loving support of family. A big thank you to Amita, my dear wife, who supported me through this journey. This book would not have seen the light of day without her unwavering support. Thank you, little Manas—my six-year-old son—for being my inspiration.

I offer my sincere gratitude to Dr. Ginger Levin, for her expert direction and advice on this book. I remember my first call with Ginger, sitting in my car in a parking lot over lunch. Ginger not only helped me get introduced to the Taylor & Francis Group, my publishers, but she was a constant source of support, guidance, and encouragement as I wrote this book.

Many thanks to everyone at Taylor & Francis Group for believing in me and bringing this to life.

And, finally, I would like to sincerely thank Brian Grafsgaard and Krissy Wolle for allowing me to interview them. The expert interviews that you see in this book are a result of them sharing their experience and knowledge. Thank you, Brian, and thank you, Krissy.

About the Author

Samir Penkar is an award-winning program and project management professional.

He is the founder of the Future of Project Management blog, the place for people, trends, and ideas on project management. Penkar has been featured in *The Wall Street Journal*, Fox 9 News, and numerous project management publications. A passionate speaker, he has presented internationally across the United States, Asia, Canada, and Africa.

Penkar has a Bachelor of Engineering in Electronics and a MBA from Mumbai University in India. He is a certified Project Management Professional (PMP) as well as a certified Scrum Master.

As a consultant, he has worked with a number of organizations in the education, insurance, fitness, manufacturing, and agro industries.

Cast of Characters

Susan Codwell: Project manager promoted to program manager
Derek Codwell: Susan's five-year-old son
Andy Codwell: Susan's husband
Monica Friedman: Attorney and a mother of two, Susan's running buddy
Steve Meyer: Musician, Susan's second running buddy
Arthur Russel: PMO director
Antonio Zubrod: CEO of FitAtWork Inc.
Murali Krishnan: Project manager
Barbara Taylor: Project manager
Harvey Larson: Technical architect
Bill Holtz: Senior business analyst
Paul Landers: Chief financial officer
Mary Beth Jensen: Vice president of National Delivery
Ronald Weinberg: Vice president of Sales
Fran Straus: Legal counsel

Disclaimer: This is a work of fiction. Names, characters, places, and incidents either are products of the author's imagination or are used fictitiously. Any resemblance to actual events or locales or persons, living or dead, is entirely coincidental.

1

The Birth of My First Program

My legs are failing, as I am trying to keep up with my running buddies, Monica Friedman and Steve Meyer. "Come on, Susan Codwell, you can do it," I am saying to myself. It's a crisp Saturday morning, and the three of us are on our usual trail run. This 10-mile trail takes us to the edge of Lake Reiley, into the woods, and past the Bear Lakes Country Golf Club.

For some reason, I cannot find my running stride today. My mind drifts to the previous day to the program review meeting with the Project Management Office (PMO) leadership. First, why would anyone schedule a critical program review at 3:30 p.m. on a Friday afternoon? A panel of stern-looking sponsors, managers, and our PMO director, Arthur Russell, are leading the charge. Arthur was my biggest fan, so I thought, but at this meeting, he seemed like my worst enemy, ready to pounce on every point I made and my attempts at trying to explain how we can get out of this program mess.

"Hey, look at that dog," shouts out Steve over his back. A happy black Labrador is hurling himself into the lake, hustling to reach the rubber bunny that his master had flung into the water. As I peek a glance sidewise, I lose my step and fall head long into the bushes.

"Oh my God, are you all right?" Monica immediately turns back. Steve is a little farther away before he realizes I have fallen. He comes rushing to my side.

"I am okay," I blurt out, brushing off the dust and leaves from my T-shirt.

"What happened? You seem very distracted today," Steve asks.

I am still sitting on the edge of the trail, and Monica hands me her water bottle. I take a few sips. "It's nothing really; maybe I did not get a good night's sleep, guys," I say, as I pick myself up. Luckily, no damage is done, except for a few bruises on my knee and arm.

"Let's turn back, it's getting warm anyway," suggests Monica. We trace our way back through the winding trail; the Labrador is still at his toy, happily wagging his tail.

The rest of my Saturday is filled with ice skating classes for Derek (my adorable five-year-old son), grocery shopping, laundry, and cooking. That leaves no place to think about my troubled program at work. This program represents a pivotal advance in my career, my chance to move up from a project manager to a higher plane.

At five o'clock my cell phone buzzes, Monica is calling.

"Hey girl! How are those bruises and are you okay?"

I love Monica's enthusiasm on the phone. "I'm okay and thanks for calling." I hesitate a little and continue, "You know, it's just that my mind was at work; we have this large program that I am in charge of and we are having some issues lately. Running was supposed to be my bliss and getaway, but maybe the work stress got to me in the morning. My arm is a little bruised, but I am going to be fine by next Saturday. I so appreciate your call, Monica."

"Hey girl, when you run, just focus on the trail and your breathing, everything else will dissolve. Well, good luck with your program, and see you Saturday then," Monica says as she clicks off.

Monica Friedman, Steve Meyer, and I met at my first marathon training program. We instantly bonded and for the past two years we have this Saturday morning run ritual. I really enjoyed the company of these two. Monica is a 34-year-old attorney, a mother of two, who still thinks partying until 4 a.m. is the best girl's night out ever. Steve Meyer is a musician; he plays the violin for the local orchestra. Even with his late night shows, Steve has rarely missed any of our Saturday morning runs. I admire Steve for following his passion of music and making it his career. Arguably one of the best artists I know, on his running T-shirt, Steve has a pepper black-colored logo studded with gold trims and a soothing green outline that he painted himself. It says: Running is Musical.

My mind drifts back to the meeting with the PMO leadership on Friday. Arthur began the meeting with some small talk about the weekend and then he abruptly announced that we may have major issues with the Fast Track Proposal Program. My company, FitAtWork Inc., started out as a corporate wellness program offering. We have branched out into consumer-driven and an entire array of wellness initiatives for organizations. We have a battery of health coaches, fitness instructors, and even doctors on the payroll. With a nationwide network of nurses, we conduct biometrics clinics at employer locations, design, and deliver wellness programs. A technology-heavy company, we have a range of online customer-facing portals, social media outreach, and smartphone apps in addition to the administrative systems that drive the internal operations. As the company has grown, we have struggled to keep up with the legacy systems, the mergers and acquisitions, and the ever-increasing demand from marketing for faster deployment of newer mobile technologies. It would be hard for someone in our company to feel like we are in the midst of an economic recession.

In my brief three years with the company, I have been promoted twice—once from a junior project manager to senior project manager and recently to a program manager. I remember walking into Arthur Russell's office for my final interview and was blown away with his office wall lined with medals from his numerous marathons, and photos of his triathlons and skiing trips. We had a great conversation and Arthur called me personally that same evening and offered me the job. I took it.

I am a solid project manager and quickly gained Arthur's trust. The very next year, he promoted me to a senior project manager. When the company began to contemplate this Fast Track Proposal Program, I volunteered to lead it, with a knot in my stomach. In the corporate wellness industry, every major corporation was installing a wellness program. The volume of Request for Proposals (RFPs) was intimidating. We could respond to only 40% of the proposal requests that we received. There was a shortage of subject matter experts, every proposal effort was a one-off effort (limited to a single time). Proposal assets were not centrally located and were difficult to obtain. And, the biggest challenge was interdepartmental coordination of sales for customer interactions, marketing for messaging, legal, Information Technology, health coaches, and support teams. The vision was to build a proposal management workflow system and content repository that would dramatically speed up the proposal creation process. This was a key factor hurting the company's growth.

If I could pull off this program, Arthur had promised a great career path for me. He even sent me to a two-day program management workshop. I had all the theory down, but never had I managed such a large initiative. My interest for program management started as a purely intellectual exercise, but was soon propelled toward a more pragmatic and now a real program.

"It's just a large project, manage it like any other project" is what I told myself. However, the skills I would need to manage this program were far more complex than anything I had ever imagined. Little did I know that when I requested that Arthur make me the program manager.

Similar to witnessing a baby arrive in this world, I had seen the Fast Track Proposal Program take birth. Its conception lay in the active brain of our CEO, Antonio Zubrod. Through its pregnancy, it was nurtured by a number of our steering committee members. It took Antonio a full five years to actually gather momentum for this start-up company. Recruiting health coaches and nurses in the initial days was an uphill task.

"To get a health coach to come and work the phones was a difficult thing to do. Not like taking a client to the gym or teaching them to swim," he recalls.

The steering committee spent a good three months debating if we should acquire an off-the-shelf solution or build our own. They decided to land somewhere in the middle. ToGetherMode Inc. was selected to provide the collaborative workspace, and then our technical development team would build the workflow, asset repository, and functions around this workspace. Arthur Russell himself drafted the initial program charter. At the time, he thought of this as a large project. There were few departments that did not in some fashion contribute to the RFPs. It soon became apparent that by installing a workflow system for RFPs, we would need multiple streams of work to bring this project to life. Another factor involved the time frame that we thought this would take. A total disruption of the RFP process was a scary thought; no one wanted to fall down below the 40% RFPs that we could respond to currently, yet to take us beyond this 40% response rate, we would require the same set of subject matter experts. Well, you add the sales team on top of all this and their time was extremely critical. Selling a wellness program to a large company is a tedious and long, drawn out process. At any point of time, a single sales manager would be creating three to four RPFs. Plus, they also had to meet their quotas and attend training for new products and services. They also were expected to update our Customer Relationship Management (CRM) system.

In the course of the year, Arthur Russell updated the initial charter to include the following points that justified why we need to treat this as a program, rather than just a project:

1. Duration: Estimated duration for the entire initiative to be rolled out and the benefits realized was over 16 months.
2. Strategic importance (large scope and many stakeholders): The initiative was of significant importance to the business of the company. Our expansion plans depended on this program being successful. It

would touch almost every major function in our organization: sales, legal, IT, marketing, health coaches, and national biometric practice.

3. Multiple funding sources: As the benefits of this initiative would spread across the groups, so was the funding for this program spread out across multiple departments. This made it mandatory for us to view this initiative at a macro level and yet track costs and benefits at the department levels.

4. Need for organizational level governance: Because of the wide reaching scope, an executive steering committee would be required to function as the governance body.

5. Ambiguous benefits quantification: Although the various teams had some idea of the benefits, it was only a guess at best. Softer benefits, such as more time available to the sales force, could translate into more sales opportunities or freeing up capacity of the marketing department would enable us to do wider outreach to potential customers or quality of the proposals would increase leading to higher sales. A precarious slope that none of the groups wanted to commit hard dollars to, but touted these reasons to justify their needs.

6. Need to track benefit realization: The benefits of this program needed to be monitored and tracked. This was a directive from Antonio Zubrod, our CEO, himself. If the benefits were not being realized, Antonio may have had a plan B, which he never shared with us.

7. Multiple work streams: We had a software collaboration package vendor, internal systems integration, a new workflow for proposal management, and loads of training. Could we have treated this like a large project? Maybe, but the multiple work streams ran in parallel, in addition, some work streams or, in other words, projects would go live very early in the program life cycle.

I had never been on a project that tracked benefit realization. Most of the times we worked crazy hours to get the system to production and then said goodbye to our baby, never to see it again. The thought of seeing our work actually realize its benefits was both exhilarating as well as scary. What if we did not deliver, what if the benefits we thought were the wrong ones? Will it affect the existing sales forecast? Will I have a job, if this thing does not deliver? These were the thoughts running through my mind in the morning, when I stumbled and fell on my run. I checked my elbow; it was tender with the bruise, but quickly drying up and healing.

My mind drifted back to the 3:30 p.m. Friday meeting with the program leadership. ToGetherMode Inc., our collaborative workspace vendor, had informed us that our proposed integration with their system would not be feasible and that we would have to change direction on the technical integration pieces. This could potentially set the collaborative work stream back by many weeks or even months. I had not seen this coming, and the impact to other projects would be very high. We had not even begun to understand the impact to the program as a whole, when what seemed like a knee-jerk reaction triggered this horrible 3:30 p.m. Friday meeting.

"You did not even have this in the program risk register," accused Arthur.

The others were much more polite. "What can we do to help?"

And, all I could muster was, "Give me until Monday to figure out a plan of action."

On this late Saturday night, I so wanted to form a cocoon to take my thoughts off the program issues.

"Mommy, let's play puzzle, puzzle." Derek came to my rescue. The rest of my Saturday was consumed with building a 100-piece dinosaur puzzle. Derek has learned to start with the border of the puzzle, and he quickly proceeded to find all the border pieces. When you have a large unknown problem in front of you, start with what you know. I decided to apply that strategy to my own problem at work. But, that would have to wait.

REFLECTIONS

1. What was that one project risk that you did not see coming? Think about why you didn't catch it sooner.
2. Which program or project had a game-changing effect on your career? Think about the challenges you faced on that project or program.
3. Have you ever been responsible for realizing the benefits of your projects/programs?
4. What is your next career move? What are you doing today to get closer to that goal?

2

What Is It Really I Do?

I love project initiation. The idea of taking something vague with a loosely defined goal and drilling it down to the deliverables, and the process of planning brings me great joy. It is one of those satisfying moments for me on any project. Once the decision was made to go ahead with the Fast Track Proposal Program, there was a hustle to get some sort of high-level roadmap and schedule drafted. That task Arthur entrusted to me and provided me with Murali Krishnan and Barbara Taylor, my project managers; Harvey Larson, technical architect; and Bill Holtz, our senior business analyst. All we had at this point in time was a brief idea document that the chief executive officer and his direct reports had compiled.

IDEA DOCUMENT

FAST TRACK PROPOSAL PROGRAM

FitAtWork, Inc. has been growing steadily for the past two years. As we plan nationwide growth, there are significant challenges both in the external marketplace as well as in operations. One key factor that is holding back growth is our ability to respond to requests for proposals in a timely and quality manner. It is estimated that we can only respond to about 40% of the requests for proposals that we receive.

Sales and marketing are the prime drivers of the proposal process. A number of challenging operational issues have been identified.

1. Lack of common proposal artifacts: every proposal seems like a new proposal and sales and marketing have to hunt for common artifacts and content.
2. Absence of a collaborative workplace: with many of the sales team being virtual compiling a proposal with the right revisions and updates is a daunting and tedious task.
3. Lack of a central place for tracking proposal progress as it moves through creation, approvals and final production. The sales team has to follow up with multiple folks to view the status of their proposal.
4. Quality of the proposals dropping: due to the administrative burden of compiling a proposal, little time is spent on creative problem solving and customizing the proposals to the client.

Due to all of these operational issues productivity of the sales staff is decreasing, quality of the proposals is deteriorating and we are missing out on potential business opportunities. In order to capitalize on as many market opportunities as we can, this proposal process need to be reformed, made efficient and speedier by order of magnitude. So far we have relied on disparate and individual heroic efforts to produce proposals, which is not a scalable model.

The goal of the Fast Track Proposal Program would be to simplify, speed up and help inter group collaboration. The potential benefits for the program are:

1. Free up at least 10% of the sales force time to focus on value added activities for customers and new market development
2. Increase our capacity to respond to proposals from 40% to 70% response rate
3. Improve the proposal content quality and delivery and increase proposal win rate
4. Make proposal pipeline easy to track and monitor
5. Increase sales by 5% as a direct result of implementing an efficient proposal process

This will be the top priority for FitAtWork Inc, for the next two years. The board has approved $2 million in capital and $4 million in expenses for the Fast Track Proposal program.

Prepared and approved by:

Antonio Zubrod, CEO & Chairman of the Board
Paul Landers, Chief Financial Officer
Mary Beth Jensen, Vice President National Delivery
Ronald Weinberg, Vice President Sales
Fran Straus, Legal Counsel

And, just as any project, there were lofty goals and ambiguous benefits. As I discovered, to my dismay, a budget had already been set and an expectation of a time line ingrained in everyone's mind. Someday I resolved to get ahead of this. My first response to being assigned this task of developing the program roadmap was to convene a meeting with our core IT team: Murali, Barbara, Harvey, and Bill.

Bill Holtz had already had some preliminary meetings with the sales, marketing, and health coaches, and subject matter experts to understand their pain points. Bill was the only business analyst that I had not worked with during my time at FitAtWork Inc. He had a reputation of being very detail-oriented and great at big picture thinking. I am both glad and nervous about having Bill on our team.

Our first meeting is set for next Tuesday at 10 a.m., and I cannot contain myself as I dive into creating a working agenda of what we want to discuss. I start scribbling on my notepad:

1. Understand the program objectives and benefits.
2. Get an update from Bill Holtz on his meetings with the subject matter experts.
3. Draft a high-level program work breakdown structure.
4. Identify high-level risks.

I had selected the end conference room, because this room has a huge whiteboard and also a window. Conference rooms with no windows sap the energy out of me. Especially for such an early brainstorming meeting, I want to be able to see the sky, trees, and grass, and it just helps me think better. That is also the reason I took up running to get out there and clear my mind.

Bill Holtz is already situated at the far end of the table, reading his pile of papers in front of him. "Hey, Bill, how's it going? I ask as I walk in.

"Not bad," Bill replies without a smile and continues his reading.

The blue whiteboard marker is totally dry. I dart out the room across to the other conference room to steal their blue marker. As I return, Murali, Barbara, and Harvey have arrived.

"The deployment last night had a problem, and the users are not able to access our health coaches' dashboard." Barbara is talking about her current reporting project. "Susan, I may have to duck out early," Barbara says as I walk in.

"Okay," I blurt out without sounding very enthusiastic. "So, welcome to the Fast Track Proposal Program everyone and thanks for making this meeting at such short notice," I began. "As you are aware, this is a strategic initiative for the entire company, and our growth depends on the success of this program. We have been tasked with developing a program roadmap and providing a structure to this program. I hope everyone had a chance to review the strategic idea document that was put together by the executive team. We want to reduce the effort and time it takes to complete these proposals, give time back to the sales folks, and increase sales as a result. Developing the program roadmap is the first item with which we have been tasked."

Murali Krishnan was the first to start the barrage of questions: "I heard that the program budget is set for $7 million. Who came up with that estimate?"

These were not questions I expected, and I sensed a feeling of hurt in that question. "I am in the same boat as you are, Murali; this budget was set by the executive team, and I have no insight into how it was created."

"What project are you going to assign me?" Now it's Barbara's turn to join in.

"At this point in time," I said, "all we have is this idea document from the executive team that I was hoping the five of us can work together and come up with a proposed program structure." I tried to bring them back to the agenda of the meeting.

"I have some ideas on how we can structure this," says Harvey coming to my rescue.

"But, let's first hear from Bill about his conversations with the subject matter experts," I propose, tentatively. Bill has already started to hand out some papers to everyone. I pick up my copy and it says—High Level Requirements. There is a neat table of requirements below it.

"I have had several meetings with the business sales, legal, marketing, and the health coaches," Bill says, "and, boy, we have a mess on our hands here. The entire proposal process is broken and it's such a one-off activity every time. Before we think about any software solution like the collaboration software, I would propose we do a business process mapping of the as-is situation. I started on that a bit, but need some more time to complete. I would say the biggest bang for the buck on this program would be the workflow implementation for proposals. Nobody has any idea where the proposal is except the field salesperson who has to coordinate across several teams. I pity those field sales folks. Just one proposal last month generated 3,000 emails with loads of attachments, not to say that they barely made the submission date. Source control for any proposal is a huge issue." Bill has a very calm voice and this brings some focus to our conversation.

"What you have with you is a list of high-level requirements that I have inferred from all the conversations," Bill went on. "We will need a cross-functional team to help us assign some level of priority to these."

We glance through the list; the precise and meticulous Bill has done an excellent job of capturing what is required. I grab the whiteboard marker and draw a huge box on the whiteboard. Inside that box I write *Fast Track Proposal Program.*

"Let's try and figure out how we can organize this," I said. This is an enterprise-wide initiative, and we started off by listing out the major chunks of work: collaboration vendor software implementation, asset management and versioning, CRM tacking for proposals, proposal workflow.

Harvey Larson inquires, "How about organizational change management? This program will touch almost every group in our enterprise. Should we treat organizational change management as a major program component?"

I add another box to our whiteboard—*Organizational Change Management.*

Bill spoke in cautious and measured sentences. "A major part of what Zubrod is expecting is that we track the benefits of this program. As a CEO, he wants to know if we are realizing the benefits of this large program implementation. Do you guys feel that benefits realization mandates a large component in this program?"

Before I had a chance to answer, Murali jumps in, "We have never tracked benefits on any of our projects, we implement the project and then the business unit has the responsibility to monitor its Return on Investment or payback. Why should we be responsible if those people did not use the system as it was intended?"

"But should we track benefit realization as part of this program, after all we are talking about a program here, not just any small project," countered Harvey.

"For the large part, these sales folks are abrasive and bossy; will they agree to being held accountable to this program team to report back benefits? I think we should leave the benefits realization to the business. Let them worry about it," added Murali.

"Granted that the benefits would be realized in the business units," I said, "but we should at least put in a framework and metrics to determine how we will measure success. Are we able to respond to 50% of the proposals? Is it that proposals are getting done faster? Are we winning more business? At least let's think through that, and add that as a program component. I believe it will take serious thought and effort to put this in place." Piece-by-piece a highly cohesive and consistent picture of the program emerged (Figure 2.1).

"We have a lot more work in front of us to define this better," I added.

Just then, someone knocked on our conference door; the next meeting group was assembling outside. We gathered our notes; I quickly took a photo of the whiteboard and darted out. Harvey caught me in the hallway.

"You know," he said, "we should really have someone from the business team in these sessions. I am thinking we can invite Ronald Weinberg to these sessions. Sales have the biggest piece of this pie, and Ronald is the one whose organization will be on the hook to realize most of these benefits. And, we have seen that, when the sales folks are engaged as team

FIGURE 2.1
Program components.

members, they feel included and then we don't end up with an 'us verses them' mindset."

"I will add him to the invite for our next meeting," I said, "but he travels so much, it's hard to get him. Maybe we can get one of his managers to come."

"You did a good job facilitating that meeting," compliments Harvey.

I felt pleased that we are off to a good start.

"Do you know what's up with Barbara?" Harvey asks as we parted. "She was awfully quiet today, not a word. I wonder why?"

What a great observation; I had completely missed that. Barbara was unusually quiet in our meeting. I make a mental note to check in with her later.

As I walk back to my desk, I run across Arthur. As I was about to give him a summary of our meeting, he said, "There is one thing I need from you by next Wednesday. Can you frame up the roles and responsibilities for your role as a program manager? Basically what does a program manger do? The executive team is not clear on your role and is wondering if we need to provide you with a business project manger to manage the effort. Just a couple of slides will be enough."

Arthur was speaking fast. "I need to rush to another meeting, but catch you later. How's it going?" Before I had a chance to respond, Arthur darted down the hallway to his next meeting.

What does a program manager do? I had to conjure up an answer by next Wednesday. Well, it was Arthur who had promoted me to program manager. Isn't it his responsibility to frame up the roles and responsibilities for this role? It was a deeper question for me, and something that had crossed my mind a couple of times since we started this program.

This question lingered on my mind as I drove home that day. National Public Radio is running a story on a new healthcare initiative in the state of Missouri; they are interviewing the program's program manager. My ears perk up; let me listen to how a program manager communicates. I turn up the volume.

"Our program will have far-reaching health benefits to the larger community. Community engagement is crucial to the success of this initiative and so we have started a number of outreach initiatives. We hope to achieve a platform of better health for our community; it is most important that these benefits be sustained over a period of time, and we are working feverishly to put in mechanisms that will ensure the longevity of the health benefits."

Engagement–outreach–platform for health–sustained benefits—these could very well apply to our Fast Track Proposal Program. When I first took up running, all of a sudden I started noticing runners everywhere—on streets, in parks, 26.2 stickers on cars, the type of shoes people wear on airplanes and grocery stores, in magazines, and on television. I was in a similar mode with program management now; everywhere I looked I have started seeing programs and program managers. But, what does a program manager do? It was a question that vexed me and required some reflection, deeper thought, and, for me, a very personal question. A run would be a great place to think about this. I reached Derek's school, and we spend the rest of the drive home talking about the cool teddy bear party at school.

Between dinner, homework, preparing for the next day, the rest of my evening zoomed by in no time. As I settled on the couch, Andy came in and flipped on the television.

"How's your day?" he asked.

"Not bad, how about yours?"

"I may have to travel a bit with this new program that we have."

"Oh, where are you going to travel to and when?"

"To Mexico, our company is building a new plant there, and I may have to go down to supervise some of the built out. It seems like it could be as early as next month."

"How long will you be gone?"

The thought of juggling work, Derek, and everything else myself suddenly seemed daunting.

"It could be two weeks to start with, but I will know more next week. How is your new program coming along, are you busy?"

Busy was not the word I would use to describe my current state. "We are just starting and have some things to figure out. Do you guys have program managers at your company?" I inquired.

"Sure we do, our plants are huge, and every new plant has a senior program manager assigned; there are so many moving parts you see."

Andy works for a large multinational that produces biodegradable stuff, such as spoons, plates, and napkins. He is in the engineering and robotics department and designs plant automation systems. Sometimes I envy his job; he can see his work physically come to life, unlike our software projects.

"With all the things that have to happen to commission a plant—land, personnel, technical knowledge, government regulations, testing, safety, construction, and vendors—our program manager's job is to make sure all the other groups work together and move our project forward."

I am amazed at Andy's concise and pointed explanation. I give him a hug, and I am too tired now to continue this work discussion, so we flip across some late night shows and then go to sleep.

Saturday creeps up on me, and I get my gear ready to meet my running buddies (Monica and Steve). It's a bit windy, as the sun peeks up from the horizon.

"Hey girl," Monica greets me with a hug. "Hope you are not going to fall down today," she teases. "I am going to kick your butts today, just see."

"We will have to see about that," Steve says. "Let's go!"

As we start down the trail, Monica tells us about her busy week with yet another case being dumped on her. "I have this crazy schedule this week, seven hours of meetings and then our partner wants to talk about our firm's growth over dinner at 6 p.m. Steve, is there an opening in your orchestra? I would just love to play an instrument and get out."

Monica is venting. We are going at an easy 14 minute a mile pace.

"You know guys, I kind of got this promotion at work. I am now the program manager for a big enterprise-wide initiative at our company," I started. "Do you have program managers in a law firm, Monica?"

"We have only lawyers at our firm, no program stuff there. But, lately there has been talk about hiring some outside project managers to manage our cases. You know, with all that coordination that we have to do. There is debate in our firm about getting some outside project management help and have us lawyers focus on the real case and technicalities of the law. Now that I think about it, most of my time goes in coordination. I would love to have someone do that for me."

"So, what do you have to do as a program manager?" Steve asks.

"Great question, Steve," I say. "I am trying to figure that out myself. But it's like I have oversight on this large initiative, I don't necessarily run any of the subprojects; there are other project managers who do that. But I make sure that the whole program comes together and that we have each project aligned with the program goals and that the benefits of the program are realized. I also have to manage dependencies across the projects and own the overall program communication." I struggle as I try to explain my role. What is it, really, I do? I can't even explain it to someone.

"Seems like you are the conductor of your program, just like our orchestra conductor," Steve comments.

"What! You are now turning Susan into a conductor. You musicians only think about music. This is corporate America, Steve," Monica says, teasing Steve.

This does not perturb Steve. "Think about it," he continues. "Audiences all over the world wonder if that fellow with an enamel wand between his two fingers has any effect on the musicians. The others make all the music, and the conductor gets the praise and salary and his photo on the CD cover. What a paradox. The person who is responsible for the music does not produce any sound. Isn't that like your role as a program manager; you don't build anything yourself, Susan, yet you are responsible to make it all come together." Steve is on a roll. "Your job as a program manager is to get the cadence of your projects right. Just like a conductor has to manage the tempo of an orchestra. You know what they teach music conductors? Just beat clearly and the musicians will take it from there. Communicate clearly and your team will take it from there. Your team members are busy creating your project; you should not distract them. No technique is as important as having a vision for the music. You should be ahead of the musicians at all times. You need to see that bend in the curve before them and help them navigate it. As a program manager, it is your job to look ahead when everyone else around you is focused on their individual notes."

I feel like I should be taking down notes, but Steve is picking up his running pace, and I try to keep up with him.

"Conductors in the olden days berated musicians. You can no longer do that. You need humility in you to lead. It does not take any energy to keep the music flowing at a fast pace, but it takes a lot to slow it down. It takes skill to change your project direction. We musicians steal a glance toward our conductor, a split-second glance. It's your job to convey to us what we need to know. It's an unspoken understanding that you develop over time. When you convey that clarity and assurance, our music and projects blossom."

Is Steve talking about music or is he talking about program management? It all fuses together. Steve is making music, and it is brilliant.

"Wow, Steve, you should like write this up and publish an article or blog or something," says Monica as she pats Steve on his back.

I am left in awe as Steve has so clearly described the range of skills and talents required of a program manager. It was as if an essential part of the puzzle has been solved for me. The confluence of music and program management produces a powerful image, one that I can relate to easily. It is a bold yet fitting comparison.

As we say goodbyes that day, I leave with a greater admiration for Steve.

Inspired and restless to write my own job description, I finally get a chance to flip open my laptop late at night. Both Derek and Andy are asleep, and I curl up in a blanket on the couch with my laptop. Quickly making the decision that I should start by identifying the words that describe the program manager's job, I begin writing. Half an hour later this is what I wrote down:

Oversight, oversee, orchestra conductor, communicator, benefits focused, manage uncertainty, strategy alignment, stakeholder engagement, governance, program cadence, risk management, resource optimization, manager of project managers, financial management, mentoring, creating accountability, leading multidiscipline teams, interface with senior management, competency in project management discipline, and integration management.

As I read these words, I began to wonder if I can distill these down to a more focused view, one that will allow me to recognize what I failed to recognize as a project manager. I pick out governance and oversight, benefits focused, and integration management as my top three. The others I feel competent to tackle as most of these I have experienced as a project manager. But the subtle nuances about governance and oversight, benefits focused, and integration management, I feel requires perception, acute listening, and focused observation. These highlight some of my blind spots as a program manager. Not that I have never had oversight of a project, or been part of the governance mechanism, neither have I never been benefits focused nor that I ignored integration management. But, now to define and own these domains provides a new perspective on my role as a program manager. My thinking needs to switch from managing scope, time, and cost to a longer-term view, and I quickly realize that I am going to be a program manager for a long time. These three things I identified will shape the centerpiece of my program management legacy. Feeling quite satisfied with myself, I insert a page break in my document and write the words *Governance and Oversight*.

When does oversight become an overhead? In the early days of my career, this question was asked of project management. It is less these days, but we do get the occasional challenge to project management hours on a project estimate. What does meaningful oversight mean? What value should program oversight bring to the projects and the overall program? Indeed, along with many other project managers, I had become exasperated at some earlier governance and oversight efforts—bloated governance bodies with a czar-like demand for unnecessary documentation and not enough accountability or transparency. This feeling of program governance not adding value to projects is something I am determined to change. What does the word governance mean? Is it as serious and important as it sounds? Governance boards are usually sandwiched between the customers and executing teams, such as the shareholders and the CEO. Projects need direction, just as Steve had explained that the musicians need direction from the conductor. The goal of a program governance board is to translate the vision of the program into performance. And, how do they do that? Set policies, guidelines, and monitor performance.

How would the governance work for our Fast Track Proposal program? I force myself back to the task of completing my job description assignment. Distraction is my constant enemy. Facebook, LinkedIn, constant email checking, and text messages are beginning to annoy me. An email longer than three paragraphs challenges my attention span these days. What's happening to me? There, I just got distracted again from completing my job description. With a good shake of my head, I focus on the next item.

A benefits-focused approach is an area that intrigues me. Never in my project management career, did I have the responsibility of overseeing benefits of projects. Build it, ship it, and move on to the next project, was my mode of operation thus far. To actually plan, understand, and report on benefits was a totally new space for me. As a program manager, this is a big responsibility, which will stay with me long after the projects have been executed. Will I be able to witness increase in sales, a reduction in the time it takes to complete proposals, and increased productivity all around as we hoped for? Benefits management is a much more ambiguous domain, one that brings the hope, exciting, yet scary and fluid. With this realization also came the urgent need to define a framework for benefits management; a plan of sorts. How will we measure these benefits, how can you attribute them to our program? What if the economy recovered, and sales increased, how would that impact our program? How am I to know that productivity is increasing, salespeople have more time for value-added work? Are we going to track value-added work? Benefits management needs some serious thought and framework, not to leave out buy-in from the stakeholders.

My last one of the list of core focus areas for a program is integration management. I have struggled with coordination across projects all along, and now to add a new dimension of having overall responsibility of the program, this will come to a forefront. We already know that we would be constrained on resource capacity, so how do we manage dependencies across projects and work streams? Should I develop a master program schedule? How do the program financials come together? If integration management was a key success factor on projects, it could very well turn out to be my downfall at a program level. Successful implementation depends on solid, proactive, meaningful, and simple integration across the program components.

Content with what I have for Arthur, I close the laptop and slip beside Andy, who is fast asleep.

REFLECTIONS

1. Have you ever asked yourself: What is it that I really do? If not, now is a good time.
2. Do you see yourself as an orchestra conductor? Are you the conductor for your programs?
3. Can you distill what you do on your job into three or four major themes?

3

Program Roadmap

My drive to work Monday morning was filled with anxiety of the unknown. Two questions lingered on my mind. First, how are we going to address the collaboration vendor's assessment that the change in direction for integration approach will cost us maybe months? The second question emanated from my brilliant Saturday night adventure with my program manager job responsibilities. Was I going overboard with the music conductor comparison? That, I would find out very soon.

My first meeting of the day is with Arthur Russel. I have already emailed him my program responsibilities and within minutes of my email, I had received this response—"Let's talk."

Now what is that supposed to mean? Why don't people just "not" reply to emails, rather than leaving me hanging precariously with these ambiguous two words—let's talk. I don't know what to make of this kind of response. I prepare myself for the worst.

"Good morning, Susan."

There is a hint of enthusiasm in Arthur's voice, a good sign, I presume. My program manager responsibilities is printed out and lying on his desk.

"So, you think program manager is like a music conductor. How interesting," Arthur does not pause. "I think it makes sense; it's a good comparison for the executive team to digest, too; conceptual, innovative, and not marred on our IT lingo jingo. I like it."

This is music to my ears. I silently thank Steve. Arthur is circling words on the printout.

"Oversight and governance, benefits management and integration management are the right areas of focus for you. I am pushing folks here for an enterprise Project Management Office, but it's a struggle. Maybe we can use this program to prove that program management should really flow across the organization and not just be restricted to the IT pieces of the program. The business, anyway, does not have the capacity or the inclination to do this type of coordination across the board. We need to get you in front of the business stakeholders very soon, and I want you to build great relations with each one of them."

"Thank you, Arthur," I said. "Do you think there is anything missing from this, that I need to focus on?"

"Not really, but I do want you to start building a program roadmap. We have a steering committee meeting coming up next Monday. Let's have a rough draft ready by Thursday."

As I will soon discover to my dismay, building a program roadmap is not a week's activity.

"By the way, Barbara Taylor was with me last Friday, and you should know that she is extremely disappointed that I selected you to lead this program," Arthur said. "She is a great project manager, but I felt she lacked some of the relationship skills that you have. Be sensitive to this fact, and if it's not working out between you two, let's talk."

This hit me like a bolt. No wonder Barbara was so quiet at our meeting last week. I need to pay closer attention to how people react. I had dismissed Barbara's silence without the slightest hint that something might be brewing there. Subtle risk factors like these could derail our program and my chance of success. I vow to be more aware of people's feelings and behaviors. Should I add this to my responsibilities of a program manager? I need to tune into these undercurrents before they end up in a tsunami.

"We have tons of work to do, Susan, and we need to have a conversation with ToGetherMode Inc. We just cannot afford months and months of delays on this integration stuff."

"I will. Have a good day." I walk out of Arthur's office with mixed feelings. My program manager responsibilities are on the right track, but what about Barbara and this vendor? My immediate need is to create this program roadmap. An integral element of a roadmap is some sense of a timeline, and, in order to get to a timeline, we need to understand project schedules for the various program components. Barbara and Murali, our two project managers, had already started groundwork on their projects. Getting to a timeline prematurely can lead to a host of new challenges. Yet, at times, it is by far the most influential component that can propel people into action. This time I invite Mary Beth Jensen, vice president of National Delivery; Ronald Weinberg, vice president of Sales; and Fran Straus, legal counsel, in addition to our regular crew of Murali, Barbara (our project managers), Harvey (technical architect), and Bill (our lead business analyst). After much juggling of the schedules, we assembled in a windowless conference room on the third floor of our office building.

All I have is the one-page executive summary and the program components' high-level view. After some general discussion, the conversation subsided, and they looked at me for direction. As usual I sprung to the whiteboard and wrote the words FAST TRACK PROPOSAL PROGRAM ROADMAP on the top in capital letters. There was almost universal agreement that asset management and versioning was one of the foundations to the new proposal process. Murali was assigned this program component to manage, and he had a tentative timeline of two months to deliver. Almost immediately the discussion moved to benefit realization. Having all the proposal assets in one searchable application would save huge amounts of browsing through older proposals, shared drives, and asking previous proposal managers for content. Murali wasn't sure how we would source these assets and how far back in time we needed to go. Ronald from sales decided to just take the last six months of assets as a starting point and took the action item for his team to assemble these for Murali. The critical success factor for this component was the adoption of this system by the sales folks. If they did not save their assets in the asset repository, we would not reap the benefits. Should we cut off access to all the shared drives or move the assets on day one? Attempting such a thing, as Ronald put it, would alienate the field sales folks. We negotiated a three-month transition period, after which the asset repository was to become the single source of proposal assets. This was more of a content gathering exercise, and Murali gladly took note of the decisions. The focus again shifted to Ronald. When would the benefits of this be realized? Ambitious, restless, and smart, Ronald penned down two months to understand the benefits of this implementation.

"Let's just poll my sales folks," Ronald said, "and get a sense of how this is working for them every month after going live."

Not the very hard measure that I was hoping for, but a start. Our attention then moved to the next program component.

"But we haven't discussed resourcing," Murali complained.

We decided to come back to it in our second level pass. Barbara had a good starting project plan for the collaboration vendor software implementation. We still had a huge question mark on the technical architecture issues and delay. Barbara has assumed that if we could resolve the technical issue in about a month, the total duration that we would require would be about five months of implementation. There was no hard dependency on the asset repository. Mary Beth Jensen from delivery and Ronald challenged this assumption, and we added our first scope change to the program. They require tight integration between the collaboration space and the asset management repository. Without the confluence of these two, they argued, it would appear as two different systems to the users. The next question was where should we add this scope? To Murali's asset management project or to Barbara's collaboration space project.

"Let's spring up an integration project across the board," said Harvey as the technical architect in him awakened.

We decide to hold off on this decision and move along. The discussion again steered toward benefits. The potential here was enormous, as was the challenge on quantifying the benefits of having a collaboration space. We had a rough idea that it took about a month for our company to churn out a proposal. No further metrics were available; none of the business folks recorded time and even if they did, they would not do so at this granular level. Ronald again volunteered his sales team to start tracking how long current proposals take through our internal process. This would provide a good benchmark to measure against.

I was doodling away at the whiteboard, trying to keep pace with the discussions.

Mary Beth declared, "Every program roadmap needs some milestones. How about we add some important milestones along the way?"

This suggestion added renewed energy levels in everyone and they all started throwing out milestones, start and end of every program component, start and end of benefits realizations phase, steering committee meetings, training deadlines, external events like annual sales conference, busy months to stay away from for deployments, warranty periods, integration points across the components. Some of these did not qualify as milestones, yet we captured everything.

It was Ronald's turn to have an ah ha moment. "As a sponsor, I would love to see how we are doing on the budget on this same timeline. Often I get these project schedules and then the budget is just a lump sum figure. It would be really helpful if we can incorporate a budget view on this roadmap."

"Let's have a table at the side with the budget and actual data," Murali suggested. "Better still, let's plot the budget along the timeline, and then we can track a rolled up budget by months."

I am glad Barbara finally spoke up. "We also need to see the benefits along this timeline."

"Whoa, now you are going to hold me to these?" Ronald teased. "But that brings up a good question," he continued. "Who are we building this program roadmap for?"

"For us, I guess. For all of us in this room to visualize how this program is going to flow," I added. "This program roadmap could be a great communications tool for almost all stakeholders. We can have different views for different audiences. Not everyone needs to see the budget, but I bet everyone is craving to see what we are up to."

My whiteboard hardly has any more space left to add anything else. "Let's have this discussion. We have about 15 minutes left, so who is the audience for this program roadmap?"

"How about we look at our stakeholder classification and determine if they would be interested in this view." Mary Beth was right on the money.

The only problem was that we did not have a stakeholder list ready. I made a note to move that up my priority list and we all agreed that this was a very productive session. I took a picture of the whiteboard with my smartphone and gathered all the stuff from the meeting room. The centerpiece of this program was evolving right in front of our eyes. I felt very proud of our roadmap; it was one of those days when you felt confident you have earned your paycheck. What a great feeling.

My next decision was to figure out what tool to use to put this roadmap to paper. After much debate, I decide to use Microsoft Project* to capture this roadmap. The new timeline feature in Microsoft Project 2010 was an excellent way to capture high-level timelines and roadmaps. I figured that as my project mangers provide me with updates, it would be much easier to adjust in a Microsoft Project Plan (mpp), rather than a Visio kind of drawing that could get out of sync very soon. The decision to use Microsoft Project also allowed me to help track milestones at a program level. In one shot I had my high-level tracking mechanism in place at least from a budget and schedule point of view. It's 6 p.m. and I am energized to go another two hours, but I decided to call it a day and spend the rest of the evening with my little one.

Andy was already home with Derek and they were playing ball in the living room. I am always afraid that they will break my grandmother's glass vase on the mantel. As I enter, they pretend that they were just kind of playing it gently, but I know these brats.

"Mommy, daddy is going to Mexico next week!" Derek runs into my arms.

"What, so soon," I exclaimed.

"The plant seems to be ahead of schedule and they want me down there next week to start the design work. I think I will be gone for two weeks."

"Two weeks, can't you go for just a week?" I was hoping for a yes that I knew would not come.

"If I go for two weeks now, I may not have to go down again for a long time. My flight leaves Tuesday."

My euphoria of the program roadmap quickly vanished as I started making dinner. "You know I had a wonderful day at work today," I started as we sat down to eat.

"Oh, really, it's not often you say that. What happened?" Andy asked.

"Mommy must have had an ice cream at work!" Derek inserted his comments.

I wish it was that easy for me; an ice cream could make Derek's day. Have we lost the wonder of simple pleasures as we have grown up? "Silly boy, I had a great meeting where we created a program roadmap."

"Oh, can I see the map? Does it have bridges?" Derek added. That really put an end to this conversation.

Having Andy gone for two weeks means that I will miss my Saturday morning runs, not to say that I also will have to juggle Derek's school and my work. What a damper to a great day. Why did Andy have to deliver his travel news today? Before I get to bed, I emailed Arthur our program roadmap and watermarked it DRAFT (Figure 3.1).

REFLECTIONS

1. What does your program roadmap look like? Think about how you went about creating it.
2. Was there a program roadmap you felt really proud of? What about it made you so proud?
3. Who is your audience for your program roadmap? Does your program roadmap provide value to them?

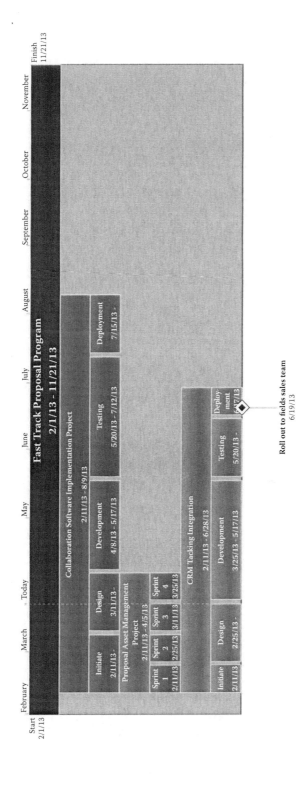

FIGURE 3.1
Program roadmap.

4

Estimates and Program Financials

It's the last Saturday before Andy's Mexico trip and I didn't want to miss my run. I am the first one to arrive at the parking lot. As I wait, I hit the button on my smartphone. These days I can hardly stay away from email. Who checks if her boss has replied to a Thursday email on a Saturday morning? I was doing it. No reply, I hit the refresh button, secretly hoping that Arthur would have replied to my program roadmap email. I hadn't seen or heard from Arthur since Thursday. Where is he? Has he seen my program roadmap?

Monica is the next to arrive.

"Flippin' horrid morning, Susan. I got a speeding ticket right before I entered the park. Can you believe it? Going 35 miles an hour in a 30 mile zone; how flippin' crazy is that?" Monica is visibly agitated.

"Cool down, Monica. Contest it if you have the time, and they will most likely reduce the fine."

"Who in their flippin' mind has the time to run after this; I am just going to mail in a check. Come, let's run."

"But, let's wait for Steve," I tried to tell her.

"Steve's not coming today. He has some kind of an early audition," Monica said. "Didn't you get the email?"

In my obsession for work, I had forgotten to check my personal email for two days. Monica has already started going down the trail. I am kind of disappointed, as I wanted to tell Steve how well Arthur had liked his music conductor and program manager analogy.

Catching up with Monica, I lock steps with her and try to match her tempo. We run for about five minutes without saying a word. Then I tell her about how Arthur Russell, our PMO director, had liked the music conductor analogy for a program manager.

"Steve's the guy, eh," Monica says.

I was about to tell her about my brilliant program roadmap exercises, when Monica suddenly asked, "So, what is your next step? How much budget do you guys have for this program? And how do you know you are going to make it?"

"Ahh, we have estimates, and we have confidence in the team to deliver," I offered, which after hearing myself seemed such a lame reply.

"You know that marathon we ran last year with Steve, we decided to break the four-hour barrier, and we did. But, boy, it looked impossible when we started training. You know how we tracked our mileage and speed every week. Do you guys monitor in the same way your budgets weekly?" Monica continued her line of questioning. "Do you have to do all that tedious accounting stuff for your program?"

"I do have to keep track of the budget," I said, "but I also have to keep track of the program benefits, which is really what is causing me to lose my sleep."

"Here's a real story from my firm," Monica volunteered. "Once my partner at the firm asked me how much time I will bill to this particular case. I said I didn't know enough about the case to estimate. 'Just give me a ballpark,' he said, which I knew would be carved in stone the moment I said it. So, I told him 500 hours. 'That's too much,' was his immediate reply. 'Then why don't you tell me how much I should bill and then maybe we can get on the same page,' I told him? You know what happened, he put down 500 hours as the estimate for that case. Be very wary of the first number you utter girl, it sticks and it sticks like bubble gum."

We are now running at under 10 minutes a mile, and it was hard to keep up the conversation. As we curved around the lake, I couldn't help but notice the spot where I had tripped about two weeks ago. My mind flipped forward to the next two weeks. Without Andy at home, it was going to be rough. Where will my program be in the next two weeks, I wondered? It was then I realized that within two weeks we would have our first report out to the steering committee. And, the first thing they are going to be looking at were financials. The estimates for the program components had been drawn up by the individual project managers; I had not even reviewed them, let alone question them. Not that I did not have faith in my project managers, Murali and Barbara, but whenever I had personally been through an estimate review, there was invariably something that I had missed. How was I going to track the overall program budget? I must set up a meeting to review the cost estimates for the component projects and with Arthur Russel to understand financial reporting expectations.

We cruised past the lake again on our way back, and two feet separated Monica and me. I was gliding behind her, letting her take the hit from the wind, while I benefitted from her draft. Perspiring profusely I could feel the harmony and glide in my run. It felt good and melted away the stress of the week and the worry of the next two weeks. It was in moments like these that I felt happy, in the zone, and ready to take on anything. As we sipped our water in the parking lot, Monica said, "You know what, I was driving a little fast this morning, and that turn I made was at about 20 miles an hour. If there was a runner coming along the path, I would have had to break really hard. I think the policeman did his job well today."

"So you are not flippin' mad at him now and, on top of it, you are giving him good karma?" I could not resist. We both laughed and gave each other high fives as we parted.

Sunday was busy with Andy packing his bags for his Mexico trip. Derek was happily running about getting stuff for Andy to put in his suitcase.

"Don't forget, you have the car oil change appointment this Thursday," Andy reminded me.

"Can I cancel it?" I objected. "The car won't stop running."

"It will only take 30 minutes." Andy is not letting up.

"I know," I said, "but to make it to the dealer, I will have to leave work by 3 p.m., pick up Derek, and then rush back to get dinner. And you know how cranky Derek gets when you are not there."

"Fine," Andy gives in. After a quiet evening, I tried to resist the temptation of logging into my email, but succumbed to the urge and flipped open my laptop. Sure enough, Arthur's email was on top of my unread emails.

"Good start to the program roadmap Susan, but we need to add more business-level milestones and also emphasize the benefits realization phases for the program components. Let's talk."

Let's talk! Why does Arthur have to end every mail with a "let's talk." I hate it. I quickly flipped to the calendar and started setting my estimate review meetings. I booked a series of meetings with Murali, Barbara, and also with the IT financial analyst, Katie. I have to get this program budget figured out.

My first meeting is with Murali Krishnan. He has the proposal asset management project. This is one tricky project, which could explode in scope pretty easily. These proposal assets have been scattered all around our corporate ether. It was no simple feat to gather them and assemble them in one place. Yet, it was one of the foundations to making the collaboration software work. The perceived notion with the sales team was that if these historic proposal assets were not part of the collaboration space, they would have a hard time starting from ground zero. There was debate about how far back we needed to go to retrieve these assets, which assets were draft and which were final versions, who would make the final call if these were to be included, and, to top it all, this was one component project that had the least budget. It was also the first project in our program to realize its benefits. Murali had suggested that we implement this project using the agile methodology. I had nothing against agile, but the steering committee needed a budget that they could sign off on, and a planned schedule that they could see. After several conversations with Murali and Arthur, we decided to adopt a hybrid approach. We would break the project up into sprints as they did in an agile project, but we would set a finite number of sprints with a stable development team. Having a stable development team made estimating easier. We had four folks planned for two months: three business analysts and one system analyst. It had $1 million allocated to it. Murali had often probed into how someone had arrived at this number, and, to the best of my knowledge, this came out of Arthur's mind. How did Arthur arrive at this, I have no idea. But, as Monica said on our run, this figure has stuck like bubble gum.

Murali had four sprints planned for this project. Their plan was to go back certain periods in time in each sprint and gather the proposal assets. It was a good way to look at things, so even if we did not get as far back as we would like to, we could set a great foundation for the collaboration project to start. I reviewed the resource plans and estimates from Murali. Nothing seemed out of place, he had the right resource rates, and the right people on the project and had documented his assumptions.

"What is your confidence level for this estimate?" I asked Murali.

To my utter surprise, Murali said, "About 30%." He continued without a pause, "This project is mostly about digging through our network drives and hunting for previous proposals, then we need to have sales bless this and find the corresponding financial data that is required for the collaboration space. First of all, I have two new business analysts on this team, their historic knowledge is mostly zero; secondly, sales has always been difficult to get hold of; thirdly, I didn't scope this out, and, talking about scope, this is blue water territory. I heard Arthur say at a meeting that we could easily go back five years and get proposal data. I am not so sure. My risk mitigation for this is to start with the most recent and work backwards. That's why I have suggested this agile approach. Let's break it up into two weeks sprints, and we shall see where we land."

This assessment from Murali gives me creeps. We don't have much contingency on this part of the program. I had been struggling to decide if each individual project within the program should carry its own contingency or if I should lump the program contingency as one bucket. There are a couple of good things about Murali's project. First is that the spend is uniform for the duration of the project; the only costs are resource costs. And, the second is that it is all expense dollars. I hate tracking capital and expense on my projects, and now I need to track it at a program level.

My next meeting is with Barbara Taylor. I am a little nervous, because this is the first time I am meeting one-on-one with Barbara since I had heard that she wanted my program manager position. Barbara has the decent $3 million collaboration software implementation. This is the heart of our program, the glue that will get all the components together, and the one with the biggest impact on benefits. Almost $3 million of Barbara's budget is for the vendor, GetTogetherNow, Inc. Barbara has been in constant communication with this vendor to understand the proposed delay in schedule.

Barbara starts with an unexpected twist, "You know, Susan, I had approached Arthur to give me this program manager job, and I hope he told you that. I am disappointed, but I will not let that affect my performance on this project. I am quite sure we will have more programs around here and that I will get my chance soon. I just wanted you to know that."

This upfront and bold statement by Barbara took me by surprise. Collecting my thoughts and voice I said, " I know, Barbara. Arthur did tell me that, and I so appreciate you being so open about it. I will do everything in my power to help you succeed. I only wish you the best, remember that." I am kind of glad Barbara came out with this herself, I would never have the courage to broach the subject with her myself.

"Let's talk about this collaboration vendor," I said. For the next 20 minutes, we discussed estimates, statement of work, resources, and contract. Barbara suggested that we keep the contingency within the projects and that I roll up the program contingency into one bucket for program-level reporting. We talked about the need for a program-level reporting dashboard for the steering committee. Barbara also suggested that we set up periodic recalibrations of the program-level budget, which I thought was a brilliant suggestion.

"Let's just put it up on the roadmap, that way everyone right up to the steering committee sees that we would perform a recalibration exercise, and they can expect to see adjustments in budget and timelines." Barbara was right on the money. We decided to do a bimonthly recalibration exercise for the program. I immediately added these milestones to the program roadmap (Figure 4.1).

The third and final component of our program was with me, the CRM (Customer Relationship Management) tracking and integration project. As the asset management and the CRM tracking and integration were running parallel, Arthur had suggested that I initiate this project and then hand it over to Murali when he frees up from his asset management project. I had almost ignored this project management responsibility; I was too busy being a program manager. As I researched this project scope, I quickly realized that unless the design for the collaboration space has been finalized, there was no way we could start this integration work. I kept the start of this project as is and decided to just delay some of the tasks to after the collaboration space design phase. This freed up some immediate bandwidth for me, and I needed all the time I could to ramp up this program. It also struck me that our program roadmap did not depict the dependencies within the program components. So I decide to update the program roadmap one more time. It was not the last time I would do this activity.

As I was updating the program roadmap, my eyes came to rest on a sheet of paper pinned to my soft board at my desk. It listed the three things I figured I must have been doing as a program manager. The three things: Governance and oversight, benefits focused, and integration management stared at me as if calling me to consider them as I updated the program roadmap. What am I missing from this roadmap? Do we have estimates for anything else I should be tracking? Of course, we have some ideas on the benefits from this program. These are estimates, too, just like our development estimates. As a program manager I am responsible for the benefits. I also need to track them just like I track costs. I immediately opened my calendar and set up a meeting with Ronald Weinberg, vice president of Sales, and Mary Beth Jensen, vice president of National Delivery Practice.

My program roadmap is looking good, and I always carry a hard copy with me at all times. My next stop was with Katie from finance. Project finance is a black box to almost everyone in the company. They have their own rules, strict policies, and sacred numbers that need a congressional committee-level approval to change. Katie showed me how the capital costs will be depreciated over five years and why we need to track expense and capital very closely. In addition, as the program may potentially flow into the next financial year, we had to break up the budget into two years. All this made me dizzy and I felt like running away from it all. Why can't

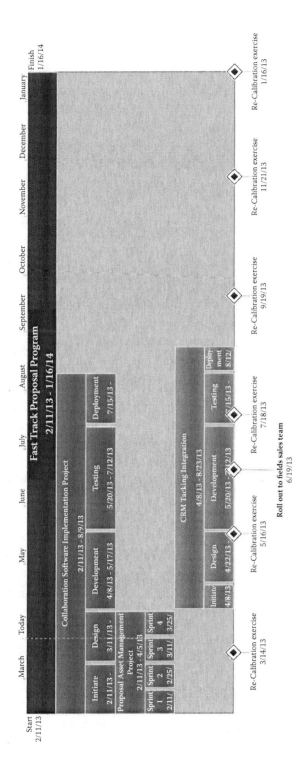

FIGURE 4.1

Program roadmap with recalibration milestones.

finance just take care of all the accounting? Anyway, what control do I really have on the budget? Which took my thoughts to, how might we control the program budget? Controlling budget means controlling costs, controlling costs means controlling resource use (in software projects, the majority of the costs are tied in people), which translates to efficient and effective implementation that requires better trained and engaged resources. I realized that I had paid almost no attention to the people in this program. We usually get handed the team and rarely get to pick the best, or the ones we want to be on our team. As I would discover to my dismay, it was a hard climb to fight for the best of the best.

I rush home to help Andy make his final preparations for his Mexico trip. We retire early on Monday night as Andy will be leaving early in the morning. I am not a morning person, not unless I want to go out for a run, that is.

"How's your new role coming along?" Andy asked unexpectedly as we were having tea. This took me quite by surprise.

"Very well indeed. I am liking it and I think we are breaking some new ground. There are a lot of unknowns still, but we are all working as hard as we can."

"Have fun, and don't take things very seriously. And, don't forget to take snacks for Derek when you go to pick him up from school, and make sure he has breakfast and he sleeps early."

"Okay, okay, I don't need schooling in parenting. Andy," I said to myself. I kiss Andy good-bye and start to get ready for the day.

REFLECTIONS

1. How do you know you are going to make your budget number? How would you answer this question?
2. How many times during a program life cycle have you recalibrated your financials? What was the result of these calibrations? Would you do them more frequently?
3. What is it about financial accounting and program accounting that you hate? Have you really thought about it?

5

Why Do It? Benefits, Benefits, Benefits

Rushing to work after dropping Derek off to school, I barely had time for a cup of coffee before my early morning meeting with Ronald Weinberg, vice president of Sales, and Mary Beth Jensen, vice president of National Delivery Practice. Our topic of discussion was program benefits. A copy of the program roadmap lay in front of them.

"How's it going, Susan?" Ronald sounded cheerful as he walked in.

"Great, let's dial in Mary Beth; she said she is travelling, but will call in," I replied. As I dial the conference call number, Ronald picked up the program roadmap.

"Hi," I spoke into the telephone. "It's Susan and Ronald here," I said.

"Hi guys, how are you doing?" Mary's voice boomed from the speaker. I quickly adjusted the volume.

"Wonderful," said Ronald. " How's your southwest region meeting? We have some good proposals in the works for that area."

"It was great. We must touch base once I am back into office. There is some great potential in this region."

I am itching to get started and, sensing that, Ronald cut off their conversation.

"Mary, let's talk about our program here. What's the agenda, Susan?"

My agenda for the meeting included:

1. Review list of planned program benefits
2. Map benefits to the program components
3. Define a measure for each benefit
4. Understand how and when benefits would be realized

"Let's start with the list of benefits," I began. "Here is the list of benefits that we have so far." I projected the list we had on the screen:

Program Benefits

1. Free up at least 10% of the sales force time to focus on value-added activities for customers and new market development
2. Increase our capacity to respond to proposals from 40 to 70% response rate
3. Improve the proposal content quality and delivery and increase proposal win rate
4. Make proposal pipeline easy to track and monitor
5. Increase sales by 5% as a direct result of implementing an efficient proposal process

"Where did you get these benefits from?" Mary Beth's voice cracked through the speaker.

"These were listed in the executive summary document that the steering committee drafted." Mary was part of that steering committee, so I was a little surprised by her question. "I just wanted to confirm with you that these are still the benefits that we envision after this program has been implemented," I remarked.

After an awkward pause of about 30 seconds, Ronald said, "At a high level, yes, these are the benefits. My problem is how are we going to measure these? Take the first one, for example, free 10% of my sales force time. I cannot measure that accurately. My sales folks don't have time to do timesheets and I am not going to ask them to. We need to change that."

"So what should we change it to?" I wait patiently for his response.

"Let's change that to: More time available to focus on value-added activities and new market development. We already capture the value-added activities in the CRM system, so that should be easy to track. And, we also capture any new market activity in the CRM system; we may have to develop some reports to help us track these. Why don't you add this reporting to one of your projects, Susan?"

I am typing furiously as Ronald had just changed our first benefit.

"Can you build these reports in the next month or so, so that we can start to get a good baseline?" Ronald asked.

"Sure, I'll check with my team." I suddenly realized that I was the acting project manager for the CRM Tracking Integration project. Shoot, I will have to raise this in priority.

"The next one is easy," Mary says. "We already have metrics on how many proposals come in the door and how many we respond to. So, that should be an easy one to tackle."

"Then let's go to the next one," I urged. Improve proposal quality and delivery and increase proposal win rate.

"I don't know about win rate," Ronald started.

"Exactly," Mary said. "How do we attribute win rate to this program? And then how do we measure the quality of our proposal? Should we have an internal review team? That will add so much more work on my team."

This was a loaded benefit; it had quality, delivery, and proposal win rate. There was an uneasy pause. How do we figure this out? As I tapped my notebook, Ronald let out an excited cry; he literally fell out of his chair.

"I got it," he shouted.

"Shh ... that was loud," Mary Beth complained on the phone.

"Here's what we should do," Ronald said excitedly. "We are trying to measure how to improve the proposal quality, right? Improving the proposal quality for whom? For our customers, or prospective customers. The best person to tell us if we have submitted a value proposal is our customer. As for win rate, that is such an absolute measure, but many times we lose to competitors not because our proposal was bad, but there is politics, there is cost cutting, there are local factors that our customers consider. So here's my idea. Why don't we just ask every customer where we have lost the bid to rate us. I am not talking about a multiple question survey, but just ask two questions: How would you rate our proposal, scale of 1 to 5 (1 = Did not meet expectations, 5 = came very close to winning), and the second question: Is there anything we could have done differently or do you have any comments?" Ronald continues without a break, "I am not talking about a formal survey through SurveyMonkey or stuff like that, although we could do that. I am talking about our field sales folks just touching base with every losing bid customer and capturing this information. We could build a simple tool to capture this data. If our ratings improve over time, we know that we are getting closer to winning, if we get feedback from the customers, we can adjust our course for the subsequent proposals. It's an easy thing to do, to measure, to implement, and we already do it on an informal basis."

"You are the man, Ronald," said Mary Beth, whose voice crackled through the speaker phone.

"We could build this simple feedback form in our CRM system itself. I'll let Barbara know to add it to her project scope," I chimed in. "There is a subjective element in the comments part of this benefit, but we can deal with that."

We debated a little about the next benefit—make proposal pipeline easy to track and monitor—and decided that this was not a discernible benefit out of this program, so we decided to cut it.

The last benefit was to increase sales by 5%. Ronald was the one to jump on this one as well.

"I think we should keep this; after this program goes into effect, we will be submitting almost double the bids as we are today, and if our win ratio holds where it is at this time, I am sure we can easily beat 5%. I would add a timeframe to this benefit, say, increase sales to 5% within six months of implementing this program."

I write the four surviving benefits on the whiteboard.

After this intense discussion, the benefits have morphed into these:

1. More time available for field sales staff to focus on value-added activities and new market development
2. Increase our capacity to respond to proposals from 40 to 70% response rate
3. Improve the proposal quality by measuring rating on losing proposals
4. Increase sales by 5% within six months of implementation

We have just 45 minutes more for this meeting, so I bring up the second agenda item.

"We now need to map these benefits to the program component," I suggested. We pick up our program roadmap and study it. "Mary, do you have the program roadmap with you?" I ask.

"Yes, I am opening it right now," said Mary Beth. We have three program components:

1. Proposal asset management project
2. Collaboration software integration project
3. CRM tracking integration project

We decide that part of the first benefit about more time being available to the field sales staff will begin to occur as soon as we start with the proposal asset management project. Hunting and finding previous proposals was a time-consuming, tedious activity for the sales staff. There were literally hundreds of folders on our shared drives and then no single naming convention or tags on which to search.

Similarly, the third benefit—improve proposal quality by measuring rating on losing proposals—had the potential to benefit us as soon as we started capturing this data. We might learn a lot more from these ratings and comments, way earlier than implementing the entire program. So, we slotted this benefit to the CRM tracking project.

The other two—increase sales by 5% and increasing capacity to respond to proposals—would only bear fruit after all the program components or projects had been implemented. So, we tagged these two benefits to the collaboration software integration project.

"Let's add these to the program roadmap," Ronald suggested. "That way every time we look at the roadmap we will not lose sight of these benefits. After all, that's what we are doing, correct?"

"Correct." Mary Beth and I responded in unison. I took the action item to update the program roadmap.

Our next agenda item was to define a measure for each of the benefit. After some discussion, we narrowed down the measures to what is shown in Table 5.1.

We ended the meeting satisfied at our benefits definition exercise. As we leave the room, Ronald says, "Susan, can you see if we can get that losing proposals rating into our CRM system? I would like to start tracking that immediately; we are only talking two fields."

"Sure, I'll see what we can do," I said. Adding two fields to the CRM software would turn out to be quite a challenge, as I was about to find out.

TABLE 5.1

Benefits Measurements Table

No.	Benefit	Measure
1	More time available for field sales staff to focus on value-added activities and new market development	Value-added activities count in CRM
2	Increase our capacity to respond to proposals from 40 to 70% response rate	Number of proposals bid on
3	Improve the proposal quality by measuring rating on losing proposals	Customer rating from losing bids and qualitative comments analysis
4	Increase sales by 5% within six months of implementation	Monthly sales figures

REFLECTIONS

1. Why? How often do you ask this question?
2. List two of your program benefits. Can you measure them?
3. Have you questioned your program benefits? Do you think you can?
4. What is it about benefits management that challenges you? Ask your stakeholders six months after your program goes live what value did it provide them? Share what you learn with your team.

6

The Prosperity Game for Governance

Barbara Taylor was tasked with trying to get the two fields that Ronald Weinberg, vice president of Sales, had requested so that we could track our losing proposals' rating and comments.

"We are the most bureaucratic, constipated, and slow company I have ever seen." Barbara is venting after multiple attempts to raise the priority level for this request with the support organization. "What's our governance structure for this program, why can't Ronald and our great PMO (Project Management Office) director, Arthur Russel, sort this out? Why am I having to break my head with these managers, trying to tell them this came from the steering committee?"

"I'll speak to Arthur, and don't get all worked up, we will sort this out," I said. Barbara is already on her way out of my cube as I finish my sentence.

Program governance is one area that we have not yet discussed as a team; we do have a series of program-level steering committee meetings set up, and they are set up once every two months. This is one area that has puzzled me in my new role as a program manager. What exactly is my role in this program governance model? Do we even have a governance model? If so, what is it? I dig through some blogs, online documents, and white papers and jot down some themes, words, and ideas on program governance, such as:

- Alignment among stakeholders
- Value
- Strategic direction
- Risk mitigation
- Shared decision making
- Manage expectations
- Manage risk

- Proper oversight
- Financial oversight
- Leadership support
- Foster agility and efficiency
- Program health checks
- Governance board
- Customers' points of view
- Alignment to mission and vision

I kind of get what this means and I don't. I wish Steve was with me to give me another of his brilliant analogies like he shared with me on our run about how the program manager is like a music conductor. Instinctively, I fire off an email to Ronald Weinberg, our most vocal and powerful sponsor:

From: Codwell, Susan
To: Weinberg, Ronald
Subject: Program Governance

Hi Ronald,

I was thinking that we should meet as a group to define our program governance structure and approach. Maybe we don't need the entire steering committee, but let me know your thoughts. Also, I would love if you have some guidance on the roles and responsibilities for our program governance board.

Thanks,

Susan

In less than five minutes, Ronald replied:

From: Weinberg, Ronald
To: Codwell, Susan
Subject: Program Governance

Susan,

Let's get the entire steering team together and I have a brilliant game that we can play to set the stage and context to the governance board.

Sent from my iPad

What game is Ronald talking about? I feel kind of guilty, as in hindsight I should have approached Arthur about this topic. After all, he is our PMO director, and he is my boss. But hindsight is always 20/20. I feel like I am developing a great rapport with Ronald. I am both excited and a little worried.

After much effort and coordination, the entire steering team gathers on a Friday afternoon. We have Arthur Russel, PMO director; my two project managers, Murali and Barbara; Paul Landers, chief financial officer; Mary Beth Jensen, vice president National Delivery; Ronald Weinberg, vice president Sales; Fran Straus, legal counsel; and Antonio Zubrod, our CEO.

As the chatter dies down, Arthur kicks off the session with a brief update on the program, and I set the stage for this meeting. We would like to define and understand program governance for the Fast Track Proposal program. Ronald is ready with his game; he has a stack of handouts for folks.

"What we are about to do here is going to be very fun, it involves money, and this exercise will help us figure out our governance model … I hope," Ronald adds in the end. He is already passing out his handouts around the table. As I get handed a copy, I am surprised at what I see.

Day 1: $100
Day 2: $200
Day 3: $400
Day 4: $800
Day 5: $1,600
Day 6: $3,200
Day 7: $6,400
Day 8: $12,800
Day 9: $25,600
Day 10: $51,200
Day 11: $102,400
Day 12: $204,800
Day 13: $409,600
Day 14: $819,200
Day 15: $1,638,400
Day 16: $3,276,800
Day 17: $6,553,600
Day 18: $13,107,200
Day 19: $26,214,400
Day 20: $52,428,800

"It starts on day 1 with $100. You must spend all of it; you cannot give it away to family and friends. Each day the sum of money doubles, and you continue to play for 20 days. Write it down; take maybe five minutes to complete your sheet." Ronald quickly goes back to his chair. I look at the Day 20 number, $52 million. My goodness, if I had $52 million, I would not be managing this program. I start on Day 1.

Day 1: $100	Buy two books
Day 2: $200	Shoes
Day 3: $400	Dress from Macy's
Day 4: $800	New refrigerator
Day 5: $1,600	Another dress by Michael Kors
Day 6: $3,200	Guccci bag
Day 7: $6,400	Redo my kitchen
Day 8: $12,800	Have a Jacuzzi in my master bathroom
Day 9: $25,600	Buy a new Honda—red
Day 10: $51,200	Down payment for new house
Day 11: $102,400	Buy an around-the-world air ticket and stay in five-star hotels
Day 12: $204,800	Buy a farm
Day 13: $409,600	Buy a new house
Day 14: $819,200	Buy a used jet
Day 15: $1,638,400	Buy a beach house in Florida
Day 16: $3,276,800	Throw a lavish party for friends and family at Las Vegas
Day 17: $6,553,600	Donate to my local school
Day 18: $13,107,200	Give it to cancer research
Day 19: $26,214,400	Take a flight into space
Day 20: $52,428,800	Donate to my city to build a museum and give it my name

I am so totally into this that I become oblivious of the others. As I look up, I see that others are in the same boat. In fact, our CEO, Antonio, is pacing up and down the room. I start to wonder what this has to do with program governance.

As the others finished buying their stuff on Day 20, we tentatively sought our neighbor to see if we can share our buying sprees.

"Okay," Ronald said as he got up again. "Are we all done with this exercise?" Seeing many nods, he continued, "Does anyone want to share what you bought on Day 1 and what you bought on Day 20?"

"I bought two books on day 1 and donated $52 million to my city to build a museum and give it my name," I volunteered.

"On day 1, I bought myself designer jeans, and on day 20, I bought an island in the middle of the Pacific Ocean," from Mary Beth.

"I bought $100 sunglasses on day 1, and on day 20, I had no idea what to do with $52 million dollars. It's beyond me to imagine," admitted Arthur.

Antonio, our CEO, went next. "On day 1, I bought myself a writing fountain pen, and on Day 20, I started a nonprofit to improve health and nutrition for kids around the world."

"Excellent," proclaimed Ronald. "The purpose of this exercise is to find out what is really important to you. Money can be meaningless after a point. What you buy with money quickly loses its significance once you start to think on a higher plane. Governance is a little like that. We, as a team, need to be thinking about the higher level goals of this program and for our company. Accountability, effectiveness, and fairness are the reasons we should care about governance. Did you realize how easy it was to get through the first five days of the exercise? Anyone can spend $1,600. But, when we jumped into the millions, you had to think hard—what is it that you really want? Things like accountability, effectiveness of the program. Of course, we need the usual stuff that consultants talk about like a governance board, regular touchpoints, and prioritization and performance measurements. But, we don't just want this formality of meetings, getting together, and policies to be a good governance body. We have assembled such a great team here, and they know how to perform. Our goal here is that every one of us at this table believes this program and its benefits. We support each other's organizations in terms of resources, finances, and priorities. I heard that Barbara was having a hard time getting my two fields added to the CRM (Customer Relationship Management) system. That's a failure on our part to not cohesively convey our priorities to our respective teams."

Ronald is now in front of the room. He says, "I am not so worried about compliance to our, say, PMO processes or to our policies. Nothing against you, Arthur. I am not so worried about consistency of processes among the various program components; let our teams decide what's the best path forward. What we, as a governance board, should worry about is really the realization of benefits. This little game that we played, you may ask what was the point of it all? This is a game that we need to repeat with program goals in mind. Think about the days in this game as our timeline for the program. As we go down the days, more and more is at stake; we are more vested in the program than the day earlier. The decisions that we make have a larger impact. So, I would really propose that we have more frequent touchpoints early in the program life cycle, and then ease out the governance calendar. In a way, we need to flip this game on its head when you think about our program. Let's start with what you can buy for $52 million, and then the $100 is a no brainer.

"We need this program team to challenge this governance board, make us cringe with their questions, hold us accountable for what we have signed up for. I want to challenge this project team to make this governance board really think hard about our program objectives and roadmap. Don't come to us just to give us a lame PowerPoint of where the program is, a pretty chart or dashboard on where we stand on the budget, which can be achieved in an email. If you are not asking us for decisions, then the governance body just becomes a stale group. All of us at this table need to raise our game to this level."

"Well said, Ronald," Antonio our CEO says. "At times, I feel really disappointed that people don't challenge our executive management enough. Just because we are your supervisors does not mean that we know it all. I like Ronald's idea of having more frequent touchpoints early on in this program, so that we are all aligned better, and understand the benefits and decisions that we need to make. And, by the way, Ronald, I love this 20-day money game. I plan to play it with my wife at home tonight."

Luckily for me Antonio addresses me directly.

"Susan, you as a program manager have a key role to play in this governance process. You need to facilitate and figure out what level of governance is required for this program, what structure fits us well. You know our culture and you need to hold the team, as well as us, accountable to our promises. You need to be always thinking about alignment to our final goal and strategy. It's a big responsibility, and the rest of the team has as much skin in the game as each one of us. Our company is at a tipping point. We can really explode in our growth and market if we get this program right. I always ask my direct reports, 'What do you need from me to be effective?' I expect that the program team will come to us and recommend things that you need to execute this program effectively."

"I will," I replied.

Antonio continues, "Just like this game, we have given this program team the budget to work with. You tell us how best to use it to get our desired benefits. Don't go and buy Gucci bags." The room burst into laughter.

I cannot contain my excitement. At the same time, somewhere deep within me, is a thought about the burden I carry as the program manager. It scares me to think that I would be responsible for letting the company down if I don't perform my job. Antonio expected us to be ecstatic about this program. This, after all, was a most ambitious program that our company was undertaking.

Coming out of this meeting, in a sense, the water around governance is muddier, but at least I feel like I know which way to head. It's like someone telling you, "Head west, and you will hit the highway." Can I live up to this responsibility? It's an exciting and scary thought.

REFLECTIONS

1. What decisions have you asked of your program governance board recently, other than asking for a higher budget?
2. Play the prosperity game with your team. What did you learn about your team when you played this game?
3. How are priorities communicated down the organization to your implementation teams? Are priorities clear in your organization? What are you doing as a program manager to communicate priorities to your team?
4. What is your role as a program manager in the governance model? Articulate your role in program governance and share it at your next team meeting.
5. Review your program benefits. Are they aligned to your program goals?

7

Program Integration Challenges

It's the second Saturday without Andy. Derek and I wake up pretty late. I feel out of place, as I am usually running on Saturday mornings. I miss my runs and miss Monica as well as Steve. Monica texted me on Friday night, just to check if I was all right. I almost thought of taking Derek in the stroller and running, but decided against dragging poor Derek out on an early Saturday morning. We have our pancake breakfast and are lounging on the sofa. I am surfing Facebook and other fashion Web sites and Derek is busy with his Ninjago game.

"Mommy, yesterday Amy's dad came to our school to tell us what he does for work," Derek says without pausing his game.

"Oh, did he? What does he do?" I asked.

"He is a children's doctor. He had a real heart-hearing machine, and I heard my own heart."

"That's called a stethoscope," I said.

"Mommy, when are you going to come to our class to tell everyone what you do?"

"Oh, I am a program manager; I help manage programs. I take care of integration, governance, and benefits management." I stop as Derek is staring at me.

"Integra…, what's that? Amy's dad said he checks to see if children are healthy. Is that what you do, too?"

"Well, it's like I check if my program is healthy," I try to explain.

"I'll ask Miss James if you can come to school," Derek sounds excited.

"Okay," I say unenthusiastically, as if to discourage Derek. How the hell am I going to explain what program management means to six-year-olds? I cannot even explain it to myself.

We spend the rest of Saturday grocery shopping, cleaning, and playing video games. Monica called in the afternoon and filled me up with their run details. I can't wait to join them next Saturday. Andy's flight is expected to arrive late evening on Sunday and Derek is all excited to see what dad brings him from Mexico.

Andy arrives at about 4:30 p.m. on Sunday. "Daddy!" shouts Derek and runs into Andy arms. I too hug Andy; I am so glad he is back. Andy immediately flings open his bag and out comes a wide, bright red, golden-trimmed Mexican hat.

"This is for you, my little one," Andy says placing it gently on Derek's head. Derek runs to the bedroom to see how he looks in the mirror. Andy also brought me some native Mexican jewelry. A big stoned necklace and matching bracelets. Andy and I talk for a while as Andy tells me all about his trip. I am glad he is back; I feel more stressed at the thought that he won't be there. Andy tells me how their plant commissioning is ahead of schedule, and how it was a great trip. They got about two days to do some sightseeing. He saw some lovely beaches, drank fresh coconut water, and had some great Mexican food.

I fill him in with my program status and some stories of Derek during the time he was away. Sunday evening the three of us enjoyed a quite dinner and retired early. I told Andy about the amazing game we played. He seemed really interested.

"$52 million by Day 20, hmm…" was his reaction.

I had an 8 a.m. meeting on Monday with Arthur.

"So, did you buy your jet yet?" Arthur teases me as I walk into his office the next morning.

I countered with, "And, did you figure out what you are going to do with your $52 million?"

"That was a really thought-provoking game, wasn't it?" said Arthur.

"It sure was. I told my husband about it yesterday and I could see he was mentally trying to play the game."

"I wanted to talk to you about a couple of things," Arthur began. "First, we have not been good here at resource management. I want you to pay close attention to how we share and manage resources across projects in this program. We don't have good data on actual resource usage. You know that. Our project reporting is spotty at the most. Let's not burden our people with excessive reporting, but find a middle ground where we can at least mitigate the risk."

"What about all the business resources that will be required to make this a success? We often leave them out and, invariably, we run into issues there," I suggested.

"Ronald and Mary Beth should be able to help us with that," said Arthur. "How are you doing? Is there anything you need from me?"

"I am doing okay," I said. "I just need some time to figure out the governance process and this cross project integration."

"How is Barbara doing? I sensed some frustration in her. Is everything okay between the two of you?" Arthur inquired.

"Yes, yes, we are fine. She was just frustrated that the request to add two fields to the CRM (Customer Relationship Management) system was tied in this bureaucratic delay. Can you help with that?"

"I have already spoken to Tony, the support manager, and they should have it done latest by next week," Arthur assured me.

"I would like you to prepare for our first steering committee update," I said, "and you have heard Ronald's view about how that should go. Let's review what we are going to present maybe a couple of days prior to the meeting." Arthur is distracted with an email that pops up in his inbox. "Arthur, there is one more thing I need from you," I said. "Can you please announce the priority of this project at the team meeting? Some folks are still not considering this as the top priority project in the organization. They seem to keep their current work at the top of the list."

"Sure, I will," Arthur replied.

For the next couple of days, I am pulled into more meetings and we discuss about interproject dependencies and resource management issues. We have a time reporting system as well as a resource forecasting system. The issue is that no one trusts the data in it—neither the resource managers, who can't figure out a soft allocation from a hard allocation; nor the developers, who hate reporting at the task level; not project managers who think of it as an overhead and can't keep up with the ever-changing resource allocations; not even Arthur, who, in fact, installed the system; and not the business, as they don't even report time. Resource management has been the thorn in our foot for such a long time that it's like we have accepted this thorn and we just limp along. Murali has a simple solution, he says that run all projects with agile, keep the team stable and allocate a resource full time to a project. That way we don't have the burden of time reporting at all. Barbara has a more traditional viewpoint: Make it mandatory to record time and forecast time, and every team member should enter their Estimate to Complete (ETC) into the system. I am not sure that either of these are the right approaches. We have three projects in this program, and if I count the total number of resources across all the projects, we are about 40 people. About 15 of these are what I call peripheral resources; we need them for specific tasks like installing servers and configuring the CRM system. So, we are left with 25 core people. Within these, we had about five resource mangers. At this starting stage, we had a fair idea about the allocation, but the problems usually arise as we dig deep into the implementation phase. It is hard to keep up with the ever-changing tasks, moving parts, and estimates. After much debate and some give-and-take, we had a plan of attack.

All resources will still log their time, and project managers will have just two tasks for the project time reporting: one task for capital and one task for expense. No tracking time at a task level below this. This made the developers happy. It also simplified the job of maintenance for project managers. The second component to this was a biweekly, one-hour resource integration meeting. All resource managers, project managers, and technical leads would participate in this meeting. We would look at each project and review the next 30 days of resource allocations. This was also a forum to discuss any resource constraints for the program as well as allow the resource managers to plan their staff. Thus, we had 1 program manager (me), 2 project managers (Murali and Barbara), 5 resource managers, and 2 technical leads, a total of 10 people.

Because we were going to meet on a biweekly basis, we also decided to discuss program-level dependencies and integration issues between the projects. I felt much more comfortable with this approach. With the rapid changing scope and shifting timelines, this was a wonderful forum to get us all connected and collaborating. It also would bring a level of transparency among the teams. We named this our integration council meeting and we would all be the council members. My job was to make sure that people truly participated in this forum. This forum could serve a larger integration function in the future; at least that was my vision. The resource managers really liked this idea. With so many projects and initiatives going on in parallel, they struggled to keep up with what's happening on the projects. With the Fast Track Proposal program being the top program in our company, everyone was eager to know what was happening and looking to contribute. My goal at this forum was to let the resource managers, project managers, and technical leads discuss integration-related issues. This was not a forum for a PowerPoint-type presentation meeting.

The first meeting went smoothly. We were just starting with the projects and resources were still transitioning from other projects. I had some trouble trying to keep our planning window to the next 30 days, as the resource managers wanted to go longer. From their point of view, it made sense, but I wanted a much shorter and immediate focus for the program. The project managers shared the status of their projects, and raised some early resource constraints. The business analysts were shared between the Proposal Asset Management and Collaboration software projects. One of the two projects would be impacted. We made the call to focus on Proposal Asset Management and allow Barbara some time to work with the collaboration software vendor. The technical leads inquired if we should have the development resources as part of this meeting. It was not really required, but we left it up to them if they wanted to invite any of their team members. The general vibe at the first meeting was very cordial and felt like we are going somewhere.

My calendar was getting filled pretty fast now with all the steering committee meetings, one-on-ones, this biweekly integration meeting, a weekly program that touched base with the project managers, my one-on-one with Arthur, and all the other meetings to which I get invited. I decided to take a look at my two-month calendar for a bird's-eye view.

Weekly: Program touch base, one-on-one with project managers
Biweekly: Integration council, one-on-one with Arthur
Monthly: Steering committee meeting
Every two months: Program recalibration meetings

Between all of these, I feel comfortable that I have covered my three major program management focus areas: governance, integration, and benefits. This makes me feel proud of what I have achieved in a short period of time. My program is beginning to bloom right in front of my eyes. It has been an exciting journey so far; kind of stressful, but seems like we are on our way. As I think about the differences between managing a single project and program of this scale, I remind myself to not lose sight of my project management fundamentals: teamwork, risk management, stakeholder management, and communications.

I realize that I have to spend a lot of time just thinking and assimilating the information that gets generated. Thinking time is a must, I soon realize, and make a decision to schedule some thinking time into my week. I block out Friday mornings to myself. I want to go out for a walk, maybe even shut myself into a conference room, and just reflect on the program progress and challenges. It's like my meditation. Steve, my running buddy, had, in fact, encouraged both Monica and me to try meditation. He said that it would clear your mind and calm your nerves. With my project managers really working the projects, I now have some liberty with my time. I block Friday mornings off through the rest of the year. And, I update the bird's-eye view of my schedule:

> Weekly: Program touch base, one-on-one with project managers, thinking time

What am I going to think about? I was not sure, but it felt right. One of the integration challenges was to think across the projects, across resources, and across the span of the entire program scope. It was like thinking about day 20 in our governance prosperity game. Our program was not an island, but was interconnected at a number of levels throughout the company. Our program must fit the organization in a seamless way. If we don't build synergy between the program and our company culture and environment, we will only create friction and delays. I realize that we have left any business-related resource discussions out of our biweekly integration council meeting. Ronald and Mary Beth are at a higher level to get involved in this biweekly integration council. However, maybe someone more hands-on from the business would be a great idea. I run this idea by Murali and Barbara, and they both agree that we often leave out the business resources and then we run into issues. Ronald and Mary Beth both assigned a manager-level person from their respective organizations to be part of our integration council. The more I think about integration, the more convinced I become that this is one of my primary responsibilities as a program manager. Not that I was not an integrator on my projects, but the level and scale of complexity that a program brings is huge. My job is to bring together these disparate projects and processes into a cohesive whole that will deliver the program benefits. Fostering collaboration and integration at all levels of the program is essential. Integration across the business groups like sales and delivery are going to be key in our program success. This IT and business mind-set has me boxing myself into the IT corner more often. As a program manager, I feel I should be owning the entire spectrum of integration. At some of the LinkedIn forums, I have often seen a project manager being described as an integrator.

So, now I am a conductor as well as in integrator. I should tell Steve that. The thought of being able to run again this Saturday cheers me up. I dig into my email and spend the rest of the day in email and running between multiple meetings.

"Susan," Bill Holtz, our lead business analyst, calls to me as I leave one of the meeting rooms. "Do you have a few minutes?"

"Uh," I hesitate.

"It won't take long I promise," he said. There is a glitter in his eyes and spring in his steps. "I attended an IIBA (International Institute of Business Analysts) dinner meeting yesterday and the speaker was discussing business architecture. The goal is to improve functional effectiveness by modeling the business to the organization's business vision and strategic goals. As I listened to the speaker, I could not help but think about our program." Bill continues to keep up with me as I walk towards my desk. "I was thinking that we should focus the business analysis at a higher level, like go across the project components and build an overall mapping. And, I believe that we should include the business as well. I even started mapping the various business entities, integration points, and their relationships. Here, take a look." Bill thrusts sheets of paper in my hand. "See, this is the entire flow of a proposal, right from the source to the final delivery."

"Bill, this seems like a great idea. Let me review this and get back to you. Thinking across the project components is a great idea, and I believe it will be valuable to the business to have this bird's-eye view. Glad you went to the dinner yesterday," I said.

"Here is the card of the speaker in case we want to invite her to provide some guidance," Bill said. "I spoke to her and she was willing to come and do a lunch and learn session at our office."

"Oh, great. Why don't you coordinate that and arrange for a lunch and learn? Make sure you get Ronald and Mary Beth in that session," I said.

"Sure," said Bill and heads down to his desk.

This short conversation with Bill makes me realize that this integration mind-set needs to percolate to our project team levels. Or should it? Should we let our project teams focus on their own project scope and not distract them with a program-level view? I didn't have the time to answer this question, as I pack my bag and head out to pick up Derek from school.

REFLECTIONS

1. What does integration mean to you? How do you manage program integration?
2. How many times during a week are you working on integration issues?
3. Write down what your weekly, biweekly, monthly, and every two months cadence looks like. Are you focusing on the right things and on the right people?
4. What mechanisms do you have in place to identify integration issues? In what forums are integration issues discussed on your programs? Is it the right forum?

8

The Long Implementation Phase

With most of my planning and definition complete, our program eased into the implementation phase. Ronald got his two fields added to the CRM (Customer Relationship Management) system, and we started to track the customer rating for losing proposals and comments. It was interesting to see that many of our losing proposals came pretty close to winning. It was Ronald's team that had to figure out why we were not winning these bids.

Murali's Proposal Asset Management project is going rather well; better than I had anticipated. His sprint approach seems to be paying off, and users are really happy that they are getting to see deliverables every two weeks. We also manage to get some additional supporting documents accumulated during this process. I personally have not used the agile management tool (see Appendix 3) on my projects, but, looking at the results Murali is delivering, I am tempted to take a hard look at it. More importantly, I have never seen the business users so engaged and on board with this approach. I guess they are reaping the benefits of a deliverable every two weeks. Who would not be happy with that? The agile approach has some other consequences that become a little harder to manage. Tracking capital and expense for this agile project is a nightmare. You are doing requirements, design, as well as development and testing all in the same sprint. Murali somehow manages to keep the financial accounting folks happy and at bay. Another benefit of this approach is that now Barbara has concrete, real, and final inputs for her collaboration software implementation project. Some of the business folks even bring up the idea that we should run the program in sprints. We discuss it briefly, but decide against it. Maybe we can try it for our next program. Another problem with the agile approach surfaced some time ago. The business analyst began feeling out of place. Murali used to have these daily touch base meetings, and the entire development team was part of them. In his sprint planning sessions, the business subject matter experts were in the room with the developers, so the developers were doing most of the interactions and clarifications. This left the business analyst feeling somewhat left out and sidelined. Murali was good at catching it and redirected some of his efforts toward integration with Barbara's project.

Barbara figured out the scheduling snafu with the collaboration vendor, and they were back on track. Barbara seemed to have gotten over her resentment that she did not get my job, and she was doing a wonderful job at managing the vendor. This was one of the centerpieces of our program. A lot depended on the success of this project for the overall program. The software vendor had an onsite team configuring the software and hardware. Barbara very early on had started conversations with the business sector about the need for resources on its side to make this a success. She had the full support of Mary Beth and Ronald, and, in fact, she met with them more often than I did these days. This was a major change to our current business process, so managing this change was a big part of Barbara's agenda, and she was always on top of it.

My third project (the one I took on as a project manager) suffered as I was focused on the program-level activities. Finally, I made the decision to transition it to Murali. His team was performing well and he seemed eager to take on another one. Murali quickly proposed that we convert this CRM tracking integration project into an agile project. We were in the middle of the design phase, so we agreed on a hybrid model where we would finish design, and then Murali can have his sprints during development and testing. Everyone seemed to be on board with this approach. I update my program roadmap one more time (Figure 8.1).

With this project off my chest, I felt better and dived furiously into my program management activities. Arthur seemed pleased with the progress, and we were just coasting along.

It is Saturday again, and I am looking forward to running with Steve and Monica.

"Hey, girl, long time no see," Monica greets me in the parking lot.

"Hey, good to see you guys. Hi Steve."

"Hi Susan. Your hubby is back from Mexico, your work life is sorted out, now this busy program manager can run with us," Steve teased.

"Oh, come on, let's run," I say.

We start off with an easy talking pace and catch up on all the weeks I have missed. Monica has a new boss, who insists that working from home is not such a great idea. Monica is cursing him as she tries to catch her

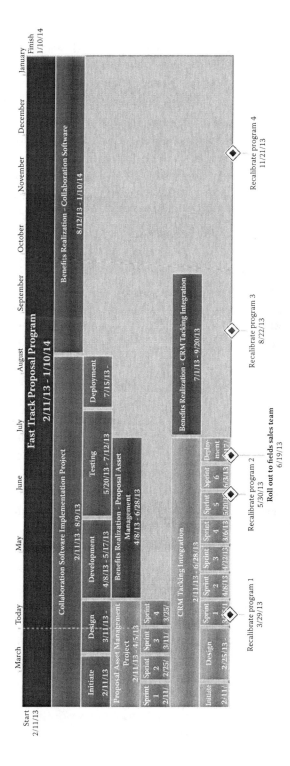

FIGURE 8.1
Program roadmap updated.

breath. Steve has landed another gig with another orchestra, and so now he has two jobs—ever-busy musician. Steve and Monica listen to how my program is progressing, and I talk nonstop for at least 15 minutes. When things are going well at work, other aspects of life also fall into place.

"Let's pick up the speed," Steve urges us. "Come on, you ladies, stop gossiping and run."

We usually run at just over 11 minutes a mile, and the goal this year is to break the 10-minute barrier. We have struggled at this for months now. I was busy with my program, Steve with his auditions, and Monica was the only one who was consistent. As we ramped up to 10 minutes a mile, I could feel my lungs gasping for air. My feet seemed okay, but my lungs were really struggling to provide the oxygen to my body. However, I kept up with Steve and Monica. Sweating profusely, we reached our turning around point. Steve was in the lead, he turned around and said, "Ladies, let's break this distance barrier today. Let's run one more mile and then turn back." What? Run one more mile? It's not like I was tired, but for months we had always turned back from this point. It was a mental mile marker for us. Somehow the run on the way back seemed much easier, and it was quicker, too. Monica and I hesitated for a moment.

"Come on, what are you thinking about, it's just one more mile," Steve is pleading.

We literally stop, I pull out my water bottle, have a shot of GU, a nutritional supplement, as if we are about to embark on a great mission.

"Come on you guys," Steve says. "What's wrong with you, I know you can both run longer distances. Why hesitate now? I'll tell you a story as we run, how about that."

"You got us with your story. It better be good, Steve, or else I am turning back," Monica says as she gives in.

We step beyond our turning back point; I feel a sense that we have broken through an invisible barrier.

"Steve, what's the story?" Monica is impatient.

"For generations, the four-minute barrier to a mile was considered a limit. It represented a physiological limit, as if the muscles could not inherently be made to move any faster or your lungs to breathe any deeper. On the afternoon of May 6, 1954, Roger Bannister broke the four-minute mile barrier. This changed the complexion of distance running forever. Within months of this, other folks broke this barrier. It was never a physical barrier. What Roger Bannister broke was not a limit, but the idea of a limit. One of the things that Roger talked about as he prepared for this was not to squander precious nervous energy before the race. It was as much training the mind as the body. The skill to reducing stress and conserving valuable energy proved to be the deciding factor to Roger's achievement. When you have long periods of doing the same thing, you begin to accept it as the norm, as the limit, as the barrier, as the insurmountable obstacle that can never be broken." Steve picked up the pace, and we kept up eager to listen.

"Think about what we broke today, our superficial barrier of going beyond our usual turning point. How you girls resisted, made excuses, there was no fence there, you know you can run longer, then why did you stop? Think about the habits or things at work you do for long phases, it is during those long phases that your faculty for new ideas and break-throughs need to be at its peak, because it is precisely the time when you need them the most."

My mind drifts back to my program. We are in the implementation phase, a long phase of the program and are we slacking. Have we lost some of the early energy that we started out with? Are we imposing a self-made barrier on ourselves?

"But, Steve, how do you prevent from going into this barrier state, as you call it?" Monica asks.

"Great question. I am glad you asked, " Steve says. "By constantly expos-ing yourselves to new perspectives, by stepping back from time to time from our distracted lives, by reflecting on our behavior and habits. Now I am not asking you to meditate like a hermit, but take this simple thing like running beyond our usual distance today. See, you are still running and we, in fact, are running at…," Steve checks his wrist band, "at 10 minutes 05 second a mile, only five seconds off our goal." Both Monica and I check our watches. Really, we are running close to 10 minutes a mile. How wonderful.

Steve continues, "The urge to improve could come from either inside or outside you. Either way, you need to be able to recognize it and act on it. You needed me today to break through this barrier. Think about it. Why did it have to be me? Why didn't one of you come up with this idea? I don't know what came over me today, but last night I was thinking about why we were not at under 10 minutes a mile after running for so long. I decided to break that mental barrier today, and thanks for following along."

"Steve, do you think we need a coach? Someone who will oversee our training, someone who will watch us perform and consult with us? Maybe that will help," Monica suggests.

"We could, or we could be each other's coaches," I said.

"That's a good idea, too," says Monica, "but I am not paying you guys anything, okay?"

I drift back to my program. Are we slacking, do we need a trigger, what is our mental barrier? How about the time I broke out of project management into program management? What was it that triggered this transition? How did I make it? How did I move out of project management? I remember the early days with Arthur when he joined as the PMO (Project Management Office) director. I had always expressed my interest in program management. I used to send him blog posts and articles for years. Any time I attended a seminar on program management, I would write up a synopsis and send it to him. I was always on his radar when it came to program management. I think that could have been a deciding factor that he picked me over Barbara Taylor. Why am I thinking about work in the middle of this fabulous run? We were about to break our 10 minute a mile barrier. I felt very proud of myself to have broken out of project management into program management. Can I ever go back?

"Great run, you guys, and thank you, guru Steve, for helping us break our freaking 10 minute a mile goal," Monica says as we sip water at the parking lot.

"Wow, this was fun. Thanks, Steve," I added.

"Welcome ladies, anytime. So are you up for breaking the 9 minute barrier next week?" We all laughed and chatted for some time before parting. As I drive home that day, I have a new kind energy flowing through me. It felt really good.

We have a quiet family weekend, and it is very quickly Monday morning. We had our integration council meeting today. This integration council was turning out to be a great hit with all the staff, especially the resource managers and technical architecture group. Many times the project teams discussed resource needs and dependencies that needed to be resolved. Most of these related to priorities for developers. Many times a priority was put to rest at these meetings. The resource managers' greatest takeaway was the understanding of the resource needs across the program. Development resources were shared across the program components and the resource managers had a hard time tracking who was working on what.

We focused on this face-to-face collaborative approach for planning. And, we focused on the next two weeks, the cadence for our integration meeting. The two-week planning horizon turned out to be a boon in disguise. Project dependencies, resource-planning needs, risks and issues were all calibrated to this two-week window. I feared at times that we would lose sight of the issues that lurked beyond these two weeks, but luckily we have not had an issue so far. Secretly, I kept going back to trying to cajole the project managers to keep expanding their planning horizon. My wish was through the end of the program, but it never happened. Where we lacked data, we made up with active and regular collaboration. Other project teams began to have these integration councils; the resource managers started requesting these for larger projects. I was pleased to see it work. The best part of our integration council was that the business team representatives also were part of our biweekly integration council. We had split the agenda into two: we discussed business-related dependencies and integration challenges for the first 30 minutes and then we let the business staff leave the meeting. Even Ronald and Mary Beth attended one time.

Another monthly occurrence was the steering committee meetings. This proved to be a challenge to meet the high standard that Ronald had set for us. The steering committee wanted to be challenged, needed recommendations and proposals to vote on, and needed leadership from the project team. All of these forced the program team to actually make decisions. It was a hard road getting folks to elevate their game and thinking. We often talked about the propensity game, and many people still carried their original sheets with their $52 million on day 20. Very soon we realized that the program roadmap was insufficient and did not convey all the information that we wanted to convey. We tried to add things, such as budget, issues, decisions that we needed from the steering committee, but it felt overloaded and never quite met the mark. Then, one day, Bill Holt again came to my desk after his adventures at one more conference. He was bubbling with excitement. Seeing him, I felt like going to more conferences. His idea was to create an infographic for the steering committee meetings. What a weird idea, I thought at first. Infographics are for marketing folks. I had seen some online infographics, but never had related this concept to projects and programs. Bill, in fact, had a rough sketch for us to consider. On an 8 ×11 sheet, Bill had doodled a rough sketch, which was to become our blueprint for the program infographic. The top part had our program roadmap splashed across the page and sections below for financials, benefits, status for each project, and then, finally, space for team recommendations, proposals, and decisions. I liked the concept, more so because I dreaded doing those bullet points presentations. This was a refresher. We decided to jazz it up a bit and discuss it at our program touch base. Bill worked with our in-house graphic designer to build a wonderful template for this program status infographic. When I was first handed this fullcolor, one-page infographic, I was blown away. Instantly, I knew that the steering committee would love this (Figure 8.2).

When we first presented this at our steering committee, I was taken aback by the level of excitement we had. Even our CEO loved the idea. With no projector to project, we found that people were more engaged in the discussions, listened better to people, they actually looked at the person speaking, rather than on a bulleted, flickering screen. There were some suggestions to add a space for business readiness and preparation. I did realize at that moment that we had taken a very IT-centric view of the program and very quickly we incorporated the business angle to our steering committee infographic. It soon became the de facto printout that everyone carried around, pinned it to their desks, and shared with any new team members who joined the program team.

Mary Beth said the best thing about our program status infographic, "This is the best one-page, stimulating, concise, and information report I have ever seen. I retain the status highlights better and find that I comprehend the program status better with this one pager."

I could see Bill beaming with pride as Mary Beth spoke. We even nominated Bill Holtz for a quarterly company award for his idea.

Nomination for Bill Holtz for quarterly company award:

> Bill Holtz proposed the idea of an infographic to present the status of the Fast Track Proposal Program. He even created a rough sketch to present the idea. When the project team welcomed his suggestion, Bill worked with the graphic artists' team to create a template for the program status infographic. This one-page program status infographic has been highly appreciated by the entire program steering committee. It has helped the program team emphasize the right information and present it in a really engaging and fun way.

We attached two copies of our program status infographics as samples.

Murali has his Proposal Asset Management sprints going very well indeed. We may even be ahead of schedule on this one. Murali also has taken over the CRM tracking integration project from me and is happily converting it to agile. I want to get to know the agile better, but just cannot find the time to do it. My program manager duties are keeping me on my toes. As I look back, I was hesitant to let Murali proceed with the agile approach, but I held back, trusting him and the team. Arthur had

Fast Track Proposal Program Status as of April 2013

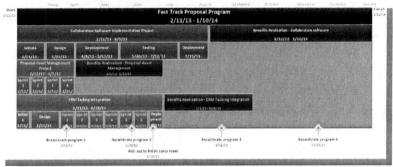

PROGRAM FINANCIALS

	Budget	Actuals To Date
$6 million	$2 million capital	$893,000
	$4 million expense	$1,100,320

PROGRAM BENEFITS

SPEND TIME ON VALUE ADDED ACTIVITES | CRM tracking

INCREASE PROPOSAL RESPONSE RATE TO 70%

IMPROVE PROPOSAL QUALITY | losing proposal rating

INCREASE SALES BY 5% IN SIX MONTHS

COLLABORATION SOFTWARE IMPLEMENTATION PROJECT

Budget/Actual: $3 million / $996,660

Schedule: 2 month delay, due to vendor integration issues. Forecast go live is planned for 10/5/2013

RISK/MITIGATION: Vendor consultant on site for configuration

NEXT STEPS:
Sales staff first look at collaboration planned
Final design review with vendor and business team
Get ready for development

PROPOSAL ASSET MANAGEMENT PROJECT

Budget/Actual: $1 million / $390,550

Schedule: Project completed

RISK/MITIGATION: None

NEXT STEPS:
Conduct lessons learned and close project

CRM TRACKING INTEGRATION PROJECT

Budget/Actual: $2 million / $606,110

Schedule: On target, planning for first sprint

RISK/MITIGATION:
Integration challenges with collaboration software could impact design / development schedule

NEXT STEPS:
Groom requirements for first sprint
Work closely with collaboration consultant on integration

PROGRAM ISSUES

1. Schedule / budget for collaboration software at risk

2. Field sales organization is requesting for smart phone access to collaboration software, this was not scoped

FORECASTS

WHAT WE NEED FROM YOU

PUT PRESSURE ON GETTOGTHER INC. to speed up configuration

EMAIL NOTE OF APPRECIATION to Proposal Asset Management team members

RECOGNIZE BILL HOLTZ for his idea in creating this program status infographic that you are reading

NEXT STEERING COMMITTEE MEETING

March 30, 2013 @ 2:00 p.m.

AGENDA: program status, decision on budget increase for collaboration software project, business team readiness review

FIGURE 8.2
Program status infographic.

pounded both Murali and me at one of his grueling PMO sessions, and only when he was satisfied, did he allow Murali to go agile.

Barbara, on the other hand, has struggled a bit with the collaboration software vendor. Some of the features that came out of the box did not function as we expected them to and so it has generated a lot of customization work. This has impacted her project budget and schedule. Working with a vendor whose project is maturing is a challenge that can have a profound impact on your success. The good part, though, is that ToGetherMode Inc. is committed to making this a model implementation and so they are going over and beyond their consulting to help us out. They even have a hands-on product administrator working with our systems engineers to fix issues and configure the system correctly. Barbara is handling them quite well. Her budget is trending higher, and her schedule has pushed out by two months. The steering committee does not seem to be too worried about this delay, but, if things go longer, I suspect they will start to worry. I know that Ronald is already concerned that our benefit realization will be delayed as we extend our timeline. As the second and third quarters near, we have our busy season when companies are selecting their wellness vendors. Barbara assures me that she has things under control. I resist the urge to dig deeper and impose myself on her project. Although I am concerned, my interference in her project will have other consequences that I have no appetite for right now. We have adjusted our timelines and benefits realization to reflect this delay.

Another larger worry for me is if the collaboration software will work as we want it to. This is a much greater risk than a two-month delay. Ronald has emphasized ease of use for his field staff right from the beginning, and the designers are working hard to make it as simple as possible. These are the two biggest risks to the program right now. At every meeting, whether a team meeting or a one-on-one, we focus deeply on the status report and ensure that we have a plan of action and are doing everything that we can to control any further delay. My program contingency will cover the overrun, but I am concerned about us exceeding even the contingency. My question to Barbara is always: "When will you know if we are going to miss our schedule and budget contingency?" And, Barbara's answer is always, "We are working on it, and new things keep coming up with the collaboration software. It is very hard for us to predict anything unless we get it all working in a development instance."

This answer really scares me. At the same time, I realize that I have given the same answer to Arthur a number of times when I was a project manager. To receive this answer now, I understand Arthur's frustrations. Murali, on the other hand, keeps telling me that I need to accept this uncertainty. We don't know what we don't know. Managing uncertainty is a core skill for a program manager. I recollect reading it somewhere. It is now manifesting itself right here, right now on my program.

My relationship with Arthur, our PMO director, seems to have taken an odd turn. Arthur now seems to share more openly his views and challenges with me than when I was a project manager. He shares his frustration at how the project managers don't even perform the basic project management tasks of periodic reporting and risk management. Was it the program that brought about this change? I do not know, but Arthur surprisingly also has shared with me his career goals. He wants to be a CIO at a larger company and thinks that this program is the feather on his cap that he has been waiting for. This really takes me by surprise and a strange, yet selfish, thought comes to my mind. Will I get to become the PMO director if Arthur leaves? I force that thought out of my head, because I need to keep my focus on our program. Its success is the stepping stone to many other career moves, but the condition is that it must be successful. The realization that many other people are banking their career and future goals on the success of this program makes me both proud and scared.

Murali wraps up his Proposal Asset Management project, and we have a minicelebration with lunch and some bowling. Now, our program is down to just two active projects. It suddenly seems a little lighter: one less project to report on, one less financial to track, and one less team to manage. As I sit back at my desk on a late Thursday afternoon, my eyes rest on this little pin up on my soft board. It has three words in my handwriting written on it below the other.

Governance
Benefits management
Integration

These were my core focus areas for program management. I feel good about two of them (governance and integration), but the third, benefits management, has yet to rear its ugly head, as I was to find out soon. I would never have understood the program management concepts if I hadn't experienced them. It's not that I wasn't aware of these before; I had read a ton of articles, blog posts, and attended too many seminars and workshops on these topics. However, nothing had prepared me for this experience. Immersion is the best teacher, and nothing can beat hands-on experience. I realize that I have risen beyond my Gucci bag.

REFLECTIONS

1. Can you build a page infographic for your project status? Think about what is most critical to communicate.
2. Have you broken down a mental barrier like the running under 10 minutes a mile barrier? What are some of your mental barriers? What are you doing today to break out of it?
3. How often do you celebrate success during the implementation phases of your program? Is it enough? Check with your teams.
4. Is your program status reporting skewed toward IT or development? What are three business items to add to your program report? Think of something other than issues and risks.

9

Benefits Management

As the flurry of implementation and "go live" activities subside, I get a breather to focus on the benefits realization phase. Murali's Proposal Asset Management project was the first one to complete. Within a month of this project being completed, we started seeing some of the benefits. Sales field staff reported that they were quickly able to find older proposal assets and it saved them some substantial time. Although this was anecdotal at this time, we accepted it as a good sign. It was harder to decipher from the CRM (Customer Relationship Management) system if the field sales staff were indeed doing more value-added activities in the time that was freed up. Ronald seemed excited from what he was hearing from his field staff. Not having a good finite and exact measure bothered me a little. It was this nagging feeling that I could never shrug off. This anecdotal evidence that sales folks were indeed saving time was not something I could digest so quickly. I am a data girl. I track my runs, my tasks, my estimates, and my issues closely. This ambiguity really did not sit well with me. Arthur kept reminding me that I need to get used to this ambiguity; perception that people are benefiting is also a benefit. Maybe, I thought, not quite convinced.

Our biggest program component had some really ugly deployment snafus. The configuration and setup of the collaboration software was not quite right in production and, as a result, the workflows automatically kept emailing the fields sales staff every four hours. Barbara spent numerous late evenings and weekends struggling with the team and vendor to set this right. However, once these issues were sorted out, the sales staff enthusiastically embraced the new collaboration space. Why would they not? It reduced clutter in their email, they were able to find their proposals quickly, understand where or who had the proposal in the process, and the best part was that they could collaboratively update a proposal. No longer did they have to rely on tedious version control and laborious consolidation of edits. At the click of a dashboard, they knew exactly where their proposal was in the process. Surprisingly, we found that the majority of the bottlenecks were with the creators of the proposal assets themselves. It was not legal or the top management approvals that were thought to be the bottlenecks. In this case, having real data on the life cycle of the proposals provided real insights into what needs improvement.

The CRM Integration project that Murali took over from me was delayed a little due to integration issues, but did not really impact the benefits too much. The field sales staff loved the fact that they could view their proposal status right within their CRM dashboards. With the program components now all live in production, I thought I would get some relief. But tracking benefits turned out to be quite a task. We had biweekly meetings with the business teams to review our benefits metrics, and some tweaks were requested in the collaboration workflows. After months of intense activity with the development teams, my time now was almost exclusively spent with the business units. Arthur had instructed me to pay close attention to benefits management. I was attending monthly sales review meetings, business outlook reviews, and, at times, Ronald's and Mary Beth's team meetings. At each of these forums, we shared our benefits tracking dashboard, discussed issues, and listened to feedback from the teams. Similar to the program status infographic, I was proud of our benefits dashboard. It was a simple Excel document for which I almost became obsessed. The benefits realization phase extended six months after the final program component was released to the organization. Almost every steering committee member carried a copy of this benefits dashboard (Table 9.1).

Arthur has assigned Murali and Barbara to other initiatives. I was the only person from the IT team to be actively involved in the program now. I sometimes felt out of place, missed the development hustle and team members. Barbara, Murali, Bill, and Harvey often stop by my desk to take a peek at the benefits dashboard and we end up chatting about how the program is doing. We had a grand party at the end of all the deployments of the individual projects. Ronald and Mary Beth made it a point to attend and showered praises on the team.

Our benefits realization phase lasted for six months. We found that the productivity of the proposal teams actually went down a little in the first two months. However, it started to creep up in the fourth month. For a while the steering committee was worried why we were not seeing the benefits sooner. Ronald assured them that some proposals were indeed taking an amazingly lesser time to create and that the quality of the

TABLE 9.1

Benefits Dashboard

Benefit	Measure	Month 1	Month 2	Month 3	Month 4	Month 5	Month 6
More time for value-added activities	Count of value-added activities in CRM	5	7	12	35	45	60
Increase proposal response rate from 40 to 70%	Number of proposals bid	82	85	83	93	101	130
Improve quality of proposals	Customer rating on losing proposals	3	3	4	3	4	4
Increase sales by 5% within six months	Monthly Sales figures	0.50%	0.75%	1%	3%	4%	6%

proposals was increasing as well. The dip in productivity was mainly due to the learning curve for the proposal teams. He asked for patience and urged the steering committee to not take any drastic steps at this point in time. Sure enough, we started to see the uptick at about four months into the benefits realization phase. And, it was a sharp uptick. Our metric for number of proposals that we could respond to jumped sharply in a month and the next month it jumped even higher. We are getting closer to our target of 70% response rate for proposals. Sales started to pick up, too, at about four months into the process. The number of value-added activities that the sales staff focused on increased dramatically in the fifth month. We actually churned out 23 white papers in that month. Our customers loved these. As we had found out, our biggest reason for losing a bid was our weak local connection to vendors and healthcare professionals in that target market. The push for building stronger relations in some targeted markets paid off, and our success rates on bids started to increase.

One of my constant struggles was to figure out how to make these benefits sustainable. Imbibing these benefits into the operations of the company was a persistent nagging risk that I could not let go. I worked tirelessly with Ronald and Mary Beth to lay the foundation of sustaining these benefits. Installing discipline among the teams to use the collaboration space as the only tool for proposal asset creation was sometimes a challenge. We struggled through slow performance issues, a corrupt database at one time, and a catastrophic downtime when the Web server crashed. But, both Ronald and Mary Beth stuck with us and we stabilized the systems. Ronald created a new role of a proposal coordinator in his organization that was a huge win for the sales team. Proposals were being tracked better, faster, and we were winning more deals. Mary Beth even went to the extent of inserting goals and targets into the annual performance appraisals for her organization. This really had the desired impact and the use of the collaboration workspace was almost universal. For the field sales folks, capturing ratings for losing bids became a passion. If fact, they used this excuse to go and meet with the customers, trying to really understand why we lost the bid. The increase in the sales had us scrambling to expand our teams and deal with some rapid expansion pains. At the steering committee meetings, I was asked to present the benefits dashboard. It was the first time in my career that I was held responsible for the benefit realization of a program. Getting into the guts of operations made me realize how pathetic a job we project managers did when we transitioned systems to operations. I vowed to pay greater attention to this issue.

REFLECTIONS

1. Think about one benefit of your program and visualize how you will absorb it into the operations of the company.
2. Create your own benefits dashboard. Can you measure your benefits?
3. Think about what you would do differently in your projects if you were held responsible for the benefits. Would you cut scope? Would you improve quality and ease of use? Would you accept last minute changes during a user acceptance cycle?

10

The Fruits of Growth

It was like my golden period; seeing the benefits come to life was a treat to me. The more I understood the business, the more I admired how well the field staff worked within our constraints and limitations of technology and process. It gave me a new admiration for what they did. Arthur was busy ramping up and selling his enterprise PMO (Project Management Office) idea to the company. He was very supportive of my involvement in the benefits phase, and, in fact, he let me get more and more close to the business side. He even handed me the responsibility of being the business and IT liaison for the National Delivery Team, Mary Beth's team. Mary was delighted. A major part of what I was focused on was to bridge the gap between IT and the business: to understand and communicate priorities to both parties, take a broader view of the portfolio of projects, and understand issues on both sides. Many times, I had asked Arthur what was next for me. Was there another program on the horizon in the company? Arthur was trying to steer me more and more toward his enterprise project management office idea. He even made me the manager for Barbara and Murali. Owning and managing people, he said, was a skill that I needed to develop. This was his gift to me for the successful program. The transition to managing Murali and Barbara was a very smooth one. We had developed great respect for each other during the program and I understood their strengths and they, in turn, trusted me. The goodwill and relations that we gained through the program was a wonderful foundation.

One late afternoon, as I sat at my desk, I plucked the sheet of paper from my pin board. It had three things written on it:

Governance and oversight
Integration management
Benefits management

Time has taken its toll on this little sheet of paper. The punched pin-hole is bigger, it has tears on the edges, and is crumpled as one day it flew off. What a journey it had been. I take a deep breath. My mind went back to the day I fell on my run, and it fast-forwarded to this day. I have covered so much ground, and moved up in my career. Steve, Monica, and I have broken the 10 minute a mile target and now are consistently running at under 9 minutes a mile. A proud feeling of achievement rushes through me.

I figure that I can now call myself a real program manager after having lived through the gusty winds of the Fast Track Proposal Program. Nothing had prepared me for this experience. You have to get your hands dirty, take that leap, and execute. Studying program management without doing it is like studying music without listening to it. Until you confront your internal fear and propose your governance model or your plan to track benefits, it is impossible to understand what it feels like. This feels like real progress to me.

Where do I want to go from here? The thought actually jolts me into reality and out of my trance. What is it that I really do? Is that the right question to ask? What is it that I really do well? Maybe I should figure that out first. The answer to that question will reveal my next career move. What does a conductor of an orchestra grow up to be? I think I need another run with Steve.

REFLECTIONS

1. How did you feel after your last project or program ended?
2. What is your career plan? Where do you see yourself in the next five years?
3. What has program management taught you? Reflect on your learning. What are those three things that you did well as a program manager? What are some areas of improvement?
4. Ask yourself, "What is it that I really do?" Then ask, "What is it that I really do *well*?"

Glossary

Agile: Agile software development is a group of software development methods based on iterative and incremental development, where requirements and solutions evolve through collaboration between self-organizing, cross-functional teams.

CRM: Customer relationship management (CRM) is a model for managing a company's interactions with current and future customers. It involves using technology to organize, automate, and synchronize sales, marketing, customer service, and technical support.

ETC: Estimate to Complete (ETC) is the estimated cost or duration required to complete the reminder of a project or program.

IIBA: International Institute of Business Analysis (IIBA) is the independent, nonprofit professional association for the growing field of business analysis.

Infographic: Information graphics or infographics are graphic visual representations of information, data or knowledge intended to present complex information quickly and clearly [1,2]. They can improve cognition by utilizing graphics to enhance the human visual system's ability to see patterns and trends.

Microsoft Project: Microsoft Project is a project management software program, developed and sold by Microsoft, which is designed to assist a project manager in developing a plan, assigning resources to tasks, tracking progress, managing the budget, and analyzing workloads.

PgMP: PMI's Program Management Professional (PgMP)® credentials recognize the advanced experience and skill of program managers. Globally recognized and demanded, the PgMP demonstrates your proven competency to oversee multiple, related projects and their resources to achieve strategic business goals.

PMI: PMI is one of the world's largest not-for-profit membership associations for the project management profession. Their professional resources and research empower more than 700,000 members, credential holders, and volunteers in nearly every country in the world to enhance their careers, improve their organizations' success, and further mature the profession.

PMO: A Project Management Office (PMO) is a group or department within a business, agency, or enterprise that defines and maintains standards for project management within the organization. The PMO strives to standardize and introduce economies of repetition in the execution of projects.

RFP: A request for proposal (RFP) is a solicitation made, often through a bidding process, by an agency or company interested in procurement of a commodity, service, or valuable asset, to potential suppliers to submit business proposals.

Scrum Master: A scrum master is the facilitator for a product development team that uses scrum, a rugby analogy for a development methodology that allows a team to self-organize and make changes quickly. The scrum master manages the process for how information is exchanged.

SME: A subject matter expert (SME) or domain expert is a person who is an expert in a particular area or topic. The term *domain expert* is frequently used in expert systems software development, and there the term always refers to the domain other than the software domain. A domain expert is a person with special knowledge or skills in a particular area of endeavor.

Sprint: A sprint is the basic unit of development in scrum. The sprint is a "time boxed" effort, i.e., it is restricted to a specific duration. The duration is fixed in advance for each sprint and is normally between one week and one month.

Standard for Program Management: The Project Management Institute's Standard for Program Management is the resource for helping program managers find the best means of achieving their goals and driving organizational success.

Appendix 1

March 2013 interview with Krissy Wolle, who is a program manage-ment professional with 20+ years of experience. You can watch the video of the entire interview at http://futureofprojectmanagement.com/fromprojectstoprograms

Samir: Hi there, this is Samir Penkar and we have with us today Krissy Wolle. She is a PgMP certified program manager with more than 20 years of experience in project and program manage-ment. And, we're going to talk to her today about program man-agement. So, thank you Krissy, and welcome.

Krissy: Thank you for having me.

Samir: Great. So, Krissy, do you want to start off by telling us, how did you get started in the program management space?

Krissy: Sure, so my first initial introduction into program management was back in about 1999 when I became a program manager of a busi-ness continuity program for a device manufacturing company that I worked for at the time. It was several projects that were unre-lated and within many different functional areas without a good overall strategy, and I brought it together under one program and became the program manager of it. Before the definition and the industry and the discipline were really, truly defined yet.

Samir: So, then you got this program together. Did the company know that it was a program? Did they just not realize that it was a program?

Krissy: No, not until I explained it to them. They didn't quite understand that they had all these related benefit projects together out there. They thought they had very disparate, projects with common threads, but not necessarily something with a common ben-efit in the end. So, they weren't being managed together, they weren't being reported together. They didn't understand how everything fit together.

Samir: Fit together, right. Now, you know, even today some people ask: "What does a project manager do?" And, program management is a relatively newer field. So, can you tell us what does a pro-gram manager do?

Krissy: Sure. A project manager is more concerned with delivering status and the objectives of a single project. Where a program manager is concerned with delivering the overall benefits of many related projects that all have components or pieces that result in the overall delivery of a specified benefit. Which is usually, a higher-level strategic objective for the company.

Samir: A lot of people ask—a lot of project managers ask—this question. What is the real difference between project and program management? There are many project managers out there who are doing a single project or managing multiple projects, but they have never managed a program. And, there are many who maybe want to graduate, I would say, to that level of program manager from a career standpoint, from a responsibility standpoint. So, what would you say are the differences in project and program management?

Krissy: I would say the key differences between project and program management, or things for someone who wants to advance from project to program to consider, would be stakeholder management. Stakeholder management in program management is much more prevalent. You don't do as much executive-level communication, presentation at the project level. You typically communicate traditionally with the project team, maybe with functional managers of different areas. Sometimes with the C level, if it's a very large project, but it's typically more focused on a program that you're doing a lot higher level communication. Because your projects likely span organizational functions and you need to communicate at that level that spans all.

So, in addition to doing communication with that group, you also have to understand what the true stakeholder management is. Not a lot of people do true stakeholder management. Where they think about what's the level of influence of this stakeholder, what do I need them to do for me, what am I worried they're going to do if I don't have their buy-in? So, you need to worry about some of the more political things with a program than you would with a project, because the span is typically greater.

Samir: Would you say that ambiguity is larger on programs?

Krissy: It depends on the program and it depends on the organization. Sometimes it can be more difficult to figure out what the overall benefit will be because there are so many different projects that contribute to that benefit. So, your scope isn't as well defined because you've got several pockets of scope versus one set of scope. So, it's a little more difficult sometimes to keep a handle on that, and to keep your stakeholders informed about that, because they don't always understand all the different pieces. They just see this overall benefit as one thing, and they want to see the results of that one thing. They don't understand there are 15 things behind it that have to happen in order to make it happen.

Samir: And, you know, benefits management is something that is a focus area for program management. So, would you say that project managers don't usually see the benefits of their projects because when they get done, they move onto another project? They're, per se, not responsible for the benefits in that sense?

Krissy: I would say the project managers are typically not held responsible or accountable for the transition of the benefit of their project, to an operational model. So, they normally get to delivery and then they hand off. If it's an IT project, they deliver, they go live, maybe they've got a short warranty period, and then they're gone.

When you're on a program, it's probable that you've got multiple projects over a longer time frame. So, that your projects that complete early in your program, even though you've delivered and you've moved onto the next one, you've got the same customer base. You're past your warranty period, and that same customer base is probably involved in projects three, four, and five. And, if there are issues with project one, you're still there and you hear about it. So, you might worry more about that operational transition then you would have with just the project because you're involved longer term.

Samir: So, for other project managers who are watching this, what advice would you give them if they want to move from project manager into a program management role? Either career wise they want to do it, or they want to manage larger projects and programs, or whatever their motive is. If a project manager is out there listening, what advice would you give them? What are the next steps? What is the first step and maybe you can start with that. What's the first next step that they should take to get to that level?

Krissy: So, the very first step for somebody that wants to move from project to program management is going to be—deliver. Deliver on the projects that you're already managing. Show your sponsors that you are capable of delivering at the lower level, and tell them that you're interested in doing more. And, chances are, they're looking for somebody that's willing, capable, and ready for that higher level. So, if you are able to deliver and show that you can excel at the lower level, it's natural, they're going to give you more responsibility.

If you think you're delivering, but you're still not getting the opportunity, think about a couple different skills. One is your stakeholder management skill. How are you doing with that? Are you actually doing true stakeholder management? Or, are you just running steering team meetings with PowerPoint slides? There's a difference. Two, make sure you're communicating. You're not only telling what the issues are and what the resolutions are, but you're making sure that everyone that's involved in the program is aware, at the level that's right for them. So, your communication vehicle for one stakeholder might be completely different than another one. You need to understand what those different communication vehicles are, and make sure you excel at them, at the project level. So, make sure you're doing well there before you try to move on to the next.

Samir: And, anything else? Would you suggest any certification, any classes, courses, in that sense; from a knowledge area, from a domain area point of view?

Krissy: Sure. Sure. So, I would definitely recommend that if you're not a PMP, you become a PMP first. I personally think that's a gradual progression. And, in addition to that, you need five years of program experience before you can become a program management professional. So, you need that PMP backing and the ability to do that role for a period of time before you can even go for that additional certification. But, definitely, get your PMP. Get knowledge in each of the areas. One that a lot of people tend to avoid, and maybe they should spend some more time there, and I think it's coming into the industry, is scope management. So, change control seems to be an area that gets focus and then it goes down for a while, gets focus just like every other area in project management. It's all cyclical.

But, for change management, it seems that project managers at times are expected to be adaptable, and that's the best feature they can have. And, that's what's going to give them the best feedback, but to be adaptable sometimes means you're constantly taking change past the date when you should. That you're accepting additional risk in your project when you should be saying no. We can't do that. We're jeopardizing quality.

You need to be able to say, "This is our scope," you want to do that change control, and we're pushing our date. If we can't push the date, you're agreeing to this quality risk. And, not a lot of project managers will say that, because they think it's going to make them appear as if they're not adaptable or not flexible. So, it's the balance with scope management and change controls that I think is one of the key differentiators between project and program management.

Samir: And, there are some organizations where there is project management happening and there's a PMO. But, there is not this concept of a program manager. And, although there could be really programs there that need to be managed at that level. If someone is in that sort of an organization, how should they go ahead and maybe propose project management to the organization or to the group. I mean, what would you suggest that they do?

Krissy: I would recommend that they research benefit delivery and benefits management, because that's the key difference between the knowledge areas in project management and program management. So, that they can try to demonstrate to their sponsors or the manager, or whoever runs that PMO, "Hey this is a distinct discipline, this is a distinct knowledge area. We need it because of, look at these pain points that we've got in our project management standards, or our process or the feedback we're getting back from our stakeholders. This is how program management would address that because there are these pieces over here in program management that aren't typically covered in project management."

Samir: So, Krissy, you've talked about a lot of things—program management, some tips for project managers on how to get into program management. One last question I have, which is: What does it mean to you to be a good program manager?

Krissy: To me, being a good program manager means that you are delivering your program according to scope, according to your budget, using the framework that's been laid out for you by the organization where you work. So, that you're working within the process and getting approval for exception when that process doesn't work, not necessarily working outside the process. Because, when you do that, you're not fixing anything. If you continually work around a broken process, that process never gets fixed. And, program managers need to worry not only about their current program, but they need to worry about the next one. Not only for them, but also for the next program manager that comes behind them. So, a good program manager is going to think about continuous improvement for the process as well as delivery of their execution for the program that they're working on.

Good program managers have good stakeholder management skills. They have excellent communication skills. Trivial things like the administration of a PPM tool or financial reporting, those are done on time. They're done accurately. They're communicated accurately. Issues and risks are well understood by everyone on the program, and they get good feedback. Now that was just a description of a perfect program manager.

They don't necessarily exist because real life in programs is not easy. There are times where you can't get everything done the way that you should, or you can't report on everything on time. But, the best is that you're trying to do what you can, you're doing it with good quality and you're following the rules as they apply.

Samir: So, Krissy, thank you. That was a good answer to what does it mean to be a good program manager. So, all of you who are listening, we talked about a lot of concepts, a lot of things. Thank Krissy for her time and for sharing with us. Thank you, Krissy. It was a pleasure.

Krissy: Thank you.

Appendix 2

February 2013 interview with the world's first PgMP (Program Management Professional), Brian Grafsgaard. You can watch the video of the entire interview at http://futureofprojectmanagement.com/from projectstoprograms

Samir: Welcome. This is Samir Penkar and today we have with us the world's first PgMP, PMI's program management professional, the first in the world to get the certification. That's Brian Grafsgaard, we have with us here today, to talk to us about program management and his thoughts about how we can all excel at program management.

Brian, for the last 2½ years has been working on *The Standard for Program Management,* 3rd edition [Project Management Institute, 2013], which will soon be released. Welcome, Brian. Do you want to start off by telling us how did you get started in this program management space?

Brian: I've never really been in an organization that had a formal career path from project management into program management or even being a functional manager or line manager in program management. I guess you could say I'm more of an accidental program manager.

I started in IT, mostly managing the development of software applications, things like that, and I just found, over time, that I was getting involved in more and more complex, larger and larger efforts, still projects, but, at some point in time, and I didn't even really recognize it at the time, my role started to shift.

I found that I was starting to lead project or team leads, and I was watching more the integration and interaction between the component projects or initiatives and was really less involved in the day-to-day project management, the formal triple constraint kind of things, which we all know is more than triple constraint now, when you count quality and risk.

Really, it was a gradual transition. I found that many times I had dual roles. Many times I was a project manager of one of the more critical component projects within a program, but was also serving in the role of a program manager, at the time, not even realizing that I was a program manager.

I'd actually been performing that role, I would say, since probably the mid-to-late '90s and, again, larger and larger and larger initiatives. I found that I was dealing with a lot broader audience too, many more stakeholders and especially stakeholders who were at increasingly higher levels in the organization.

It wasn't uncommon to be dealing with senior executives. I didn't even realize that there was a name for it, that it was actually program management that I was doing, because, with the lack of a career path, I was more of a senior project manager. I had large teams. I was actually leading other project managers, eventually. I did notice that my role was shifting, but didn't quite have a name for it (yet).

Then, one day, in 2006, I got an email from PMI and they were looking for volunteers to apply to help write the new exam for PgMP credential. When I looked at the list (of qualifications), I just dismissed it at first, then came back to it again. I looked at the list of criteria and it perfectly defined what I was doing. It was called *Program Management*.

Eventually, I did apply for that and that's really what started me on the course of becoming the first PgMP in the world, as well. I was actually one of, at the time, 13 people accepted in North America to help write the exam. I was in the very first group that helped do that.

What was interesting about that is I really didn't know that what I was doing was called *program management*. To me, I was still dabbling in project management, but also had this very different role in program management. Then, it just really dawned on me, well, I've been doing this for quite some time and I really didn't call it anything.

Samir: For those of you who are listening, do read those emails, you never know what can land in your inbox and you could be someone like Brian, the first certified person in the world for the next certification.

Anyway, program management has been around for a while now and, as project managers, there are so many questions, I think, even today, about what does a program manager actually do. So, what *does* a program manager actually do?

Brian: I think I'll frame the response in terms of the performance domains that are in the new standard for program management, the third edition, that I helped develop over the last couple of years.

If you look at a program manager, there are really five domains. In some of these, there are parallels with project management, but it's almost like the parallels are almost the project management on steroids kind of thing.

Yet, that said, I would say it is also a very distinct discipline. It is not project management. The five domains are governance, strategy, benefits, stakeholder, and we call it in the third edition—*stakeholder engagement*. We call it stakeholder engagement for the reason that, as a program manager, you're typically dealing with much higher levels in the organization.

Many of them, significantly outrank you as well, and they tend to not like to be managed. They're engaged versus managed. We actually did change the name of that actual domain in the third edition of the standard, because we recognized that, and we'd gotten feedback on that.

Then, of course, the fifth domain is just managing the program life cycle itself. I will briefly touch on each of those … Governance is … You do that in project management but, typically, again, at a program management level, it's typically at a much higher level.

You're impacting more breadth of the organization, sometimes even more depth, and you have to develop a governance structure because you're typically playing with a lot of money, as well, significantly more money than you would in a project, because if you think about a program, it's comprised of component projects.

If you look at the budgets for those in aggregate, it can be a fairly staggering number. It tends to draw the attention of very senior-level people in the organization.

The governance structure is just establishing, typically, what you'd do in project management, but you'll need a steering committee involved. You'll need some kind of program management office or even an enterprise program management office to help provide the support that you need to make sure that your program is delivering as expected and according to your baseline.

That baseline, too, in a program, I think is one of the key distinctions. As a project manager; most of us want to control change. You have a plan; you're supposed to develop on time, on budget, with the scope agreed-upon, at the proper level of quality.

As a program manager, I've found that you have to embrace change. Your program will change over time. You may have the greatest plan in the world and a perfect roadmap and something will change in the organization. I've never had a program that ended the way I thought it would when I started.

There may have been a shift in the strategy for the organization, and that's happened multiple times to me, where something that was in the roadmap may have impacted only one component project, maybe several. Something changed and it just wasn't as important as something over here, anymore.

Ultimately, it didn't mean that the program failed. Actually, the program was more successful because we were able to adapt and, I would say in a nutshell, that's it with program management, it's maintaining alignment. That's where we get into the strategy part.

A significant initiative, like a program will naturally align with that strategy. When I was talking about embracing change, that strategy may change. You may have a two-year program. You may have a 10-year program.

In terms of strategy and just duration, its kind of another differentiator with projects, in that we all know by definition they're a temporary endeavor. They have a definitive beginning and an end. That's not always the case with programs. I think, being on the core team for *The Standard for Program Management*, 3rd edition, I got to see some very high level, very, very good program managers.

We had people from the Centers for Disease Control. We had people from NASA. We had people from Boeing and, it really changed my definition of a program. But, ultimately, even in our kickoff meeting for *The Standard for Program Management*, we spent nearly two hours debating what is a program, because that was the baseline for the whole standard. We all had to have a common understanding.

Now, my programs, they always did have a finite duration. They were transitioned and became operational and we'd go do something else, but think of something like the space shuttle. That was a nearly two decade-long program. It didn't just happen and they didn't just get to walk away from it until they actually did decommission the program.

You think about the Centers for Disease Control. The program is not only to build the capability. Let's say it's smoking cessation or to eradicate tuberculosis or something like that. They're developing a capability to do that out in the world, but, yet, it is an ongoing, really never-ending, program.

It never stops. In a nutshell, I would say, in that domain, it is really about maintaining alignment, adapting your program and being able to constantly, I'd say, engage the stakeholders, because that's where the program is coming from.

Your stakeholders are really the ones who are helping you define what the objectives of the program are, how does it fit with the organization, stakeholder being that third performance domain. All the things that you would do as a project manager are, I would say, amplified, with the stakeholder management.

Again, typically in higher levels of the organization, broader, because, if you look again at the projects, every project has stakeholders, people who are impacted, have an interest, have a positive or negative attitude towards it. At the program level, you take all of that in aggregate and then add another audience besides them.

I think the best way to characterize it is, everything becomes almost 360°. You're looking at everything going on in the constituent projects. Those stakeholders are your stakeholders.

You have a broader audience, perhaps even external to the organization. You're getting information or engaging stakeholders from the top. It comes full circle in a sense, just as the risks and issues are and being able to manage the interdependencies between those.

The key distinction, as you know Samir, is you're not really managing the projects like a project manager. It's more important to manage the interdependencies between them.

I think one of the other differentiators of a program, too, is just the level of ambiguity, ambiguity meaning the "what," what are you delivering? Sometimes, with a program, you really don't know yet. You don't know how that benefit will manifest itself or what is going to be delivered.

It takes experimentation and it can take experimentation through component projects. That could be the very intent at the beginning of a program, to have projects to quickly learn, understand what works and what doesn't. It's a way that I would say a lot of innovation happens.

You may have a project that fails, and that's fine. It may not have worked out. It may not have provided the intended results. That doesn't mean the program or initiative has to fail. You take the lessons learned from that. You redefine your hypothesis or theory and you start again.

Typically, when you have a project you need at least to know the "what." You have to know the scope to build your plan around. A lot of times with a program, it's not really well defined. Ultimately, it leads us into the fifth domain—benefits. That, to me, is the key distinction between project and program management.

With project management, you have scope, you have schedule, and you have the cost. With the program, your eye always has to be on the prize. As a program manager, you are always looking at the benefits it will deliver and preparing the organization for that.

It doesn't happen by itself. One of the key tools that you'll do is develop a program roadmap. That's kind of like that Gantt chart that we were talking about.

An understanding where you might realize some incremental benefits... The key differentiator, I would say, between project and program management is really the benefits realization planning, that roadmap, and always keeping your eye on the benefits. That's the first thing you ask, as things are changing, what is the benefit of that? Or, if this component project is delivered late, what does that do to my benefits? Am I too late to market? Is it all mute now?

It's really kind of keeping your eye on all of that, the alignment, but, again, with looking at it through the benefits lens, how it's going to benefit those within the organization, with the new capabilities, but the ultimate benefit is perhaps outside the organization, especially when you think of the examples with the Centers for Disease Control. There's benefit internally to develop the capability to manage and sustain the program, but the real benefit is outside the organization.

Samir: That's true. That's a great explanation of what a program manager does and, also, it clarifies the distinction between the projects and the programs. You know, it's a whole different plane that you're really operating on, in that sense. For the program managers who are watching, what would you say are the three critical success factors for a program? What are maybe those top three things that these program managers need to focus on so that their program ends up in a successful program?

Brian: I would say number one, and we touched on it when we're talking about the five domains, but it's that benefits management lens. That, to me, is not only the number one differentiator, but the number one critical success factor.

Benefits management really has a life cycle of its own, just like a project has a life cycle and we talked about the program life cycle. Benefits management has a life cycle. One of the critical elements of that is you're identifying the benefits for the program. Early on, you're planning the benefits. Hopefully, you can realize some incremental benefits along the way, like we talked about.

You're actually delivering the benefits through the program life cycle, through those component projects, and the outcomes of those, but the key distinction, and what really is the ultimate critical success factor is, those benefits are successfully transitioned to their operational state.

Finally, they have to be sustained. That's actually one of the new things in the program management standard. We recognize that there is an ongoing phase after the program called *benefit sustainment*. The program manager is responsible for positioning the organization to sustain the benefits long term. That to me, is the number one critical success factor and you could almost say that it's having a real focus on organizational change management, both internal and external.

It's hard for me to pick three because I would like to pick one for each of the domains, because they're all equally important. I would say, second is, and it's all kind of tied together, the stakeholder management, but also managing those interdependencies, so, really kind of managing across the life cycle, helping your project managers.

I think the other thing, too, it's a third and it really kind of ties into the strategy and alignment and the governance, but it's also being able to deal with the ambiguity that is often inherent with the program—and the shifting sands. It's constantly shifting and, really, in a sense, that's the beauty of program management, that as a program manager, like we talked about, you're embracing that change. You might have the best thought-out plan and dependencies between the component projects and one might be late. One may be canceled. The strategy may shift. You decide, we're going to pull that one out. We're going to plug this in. Or, we're going to move everything forward. Or, you might have to insert one right in the middle that you had no idea that you had to do.

Again, that's kind of the beauty of it and it's a bit ambiguous by its nature and a bit uncertain.

Of course, the uncertainty kind of ties into the risk management that we talked about. As David Hilson says, "Risk is measurable uncertainty. Uncertainty is unmeasureable risk." Ultimately, it all kind of comes back to the risk management, but ambiguity is something entirely different. Just being able to always adapt, I think, is very critical.

Samir: Now you know the critical success factors for a successful program. But, as project managers, and many of the project managers who are watching, for them, Brian, what would you advise them if they want, from their career standpoint, to move from a project manager level into program management, what advice would you give these folks?

Brian: Samir, that is another great question. Like I was telling you when I was discussing my career, I just kind of grew into the role, without even really realizing it had a different name. I knew the role was different.

I would say, first and foremost, if you're a project manager now and want to become a program manager, is study program management. Understand first what it is. Make sure you really understand the distinction between project and program management and then you'll be able to do that.

Ultimately, that's been my experience, since again, I've been in organizations without a formal career path from project to program manager. Fortunately, that's changing. When I say you may have to make your own opportunities, part of those, you may have to sell your organization on the value-add of program management and actually define what that is and how the organization will benefit from that. That really ties back to really understanding the distinction between project and program management.

I would say, there are really three broad categories of programs. One is a strategic program. Those are the way programs typically are born. They start from the strategic level on down. You may be involved in, you know, one of the key component projects of that or you may be involved in just helping flush out what the program is.

They may not even realize it's a program at the beginning. For most organizations, everything looks like a project, especially for those that don't do program management, even though they do. You may be involved in the front end of a larger initiative and, if you understand the distinction, understand the value that program management provides; you may be able to help steer in that direction.

Maybe, it may develop into a more elevated, so to speak, role for you. You may be responsible for more than just your component project. Maybe you're assisting a director or a program director or something and watching the interdependencies.

The second type is broadly categorized, would be a compliance program, so, not necessarily strategic in nature, but something you have to do anyway. A lot of times with compliance programs, it truly is a people, process, and technology kind of initiative.

If it's compliance, although there may be very strict technical requirements, ultimately, compliance programs are about organizational change and behavior change. That actually lends a perfect opportunity, in my view, especially in my experience with compliance programs, to actually take a lot of disparate projects or efforts and tie them together into a more cohesive unit.

Typically, you might be dealing with the same stakeholders, over and over, whether it's a technical solution or whether it's a behavior change that you're trying to introduce, but if you see the commonalities of something that would be better managed as a program, you may just have to volunteer to say, "I will do that."

Now, the third type is a great way to get involved in program management as a project manager. The third type would be called *emergent program*. An emergent program is one that there may be strategic initiatives that have all come out of the strategic planning process and through the portfolio, but the linkage between them isn't recognized. Create the opportunity for yourself. Make it visible to your management that you're doing that, that you've recognized this opportunity, that the benefits may be greater to manage or coordinate these as a whole, and perform that role.

The only caution there, of course, is that you can't do that and sacrifice your own project. Like we talked about before, something's going to give. You really have to make sure you're balancing that and, first and foremost, performing at your job that you're supposed to be doing as a project manager.

If you have the capacity to take on that additional role and really be an ambassador within the organization for what program management is and the benefits it can provide and the value you can provide in doing that role.

Samir: You've discussed a lot about program management, what it is, the relation between projects and programs, critical success factors, how do program managers step into the project manager role. I have one last question for you. What does it mean to you to be a good program manager?

Brian: Great question. The answer to that could probably be inferred from all the things that we talked about, but I think to be a good program manager that blocking and tackling is very important. The tactical things that it takes to do a project, it takes tactical things to do a program too.

I think the differentiator and what makes a good program manager is just really staying engaged, engaging the stakeholders, letting them know, just taking care of the organizational change management that is inevitable in any program of any kind and preparing the organization for the change that's coming along the way, so those benefits can be sustained.

Many programs impact almost everybody in the organization and just making sure that you're staying engaged and always keeping in mind that in that pool of stakeholders are your project managers to be very, very supportive, to be proactive, to be helping them, not micromanaging, to just understand the challenges that they face.

I often thought, when I was project manager, that it was the toughest job in the organization and, in many respects, it is. Program management is tough, too, but also, just like project management, it can be very, very rewarding.

I also say, just thinking strategically and be willing to embrace change. It will change, undoubtedly. With projects, we try to control change and justifiably so, but, with programs, those projects are the building blocks. There may not be a reason for a particular project anymore. It doesn't mean anyone's failed.

It means that things have changed. You have to adapt and I think that's probably one of the biggest keys is being able to cross that bridge and say, "I'm not going to try to control change. I'm going to embrace it. I'm going to maintain alignment," and, ultimately, it's just a natural thing.

When you've been a program manager for a while, it's just a natural thing that it's going to change. You have to try to at least be proactive about it and you do that by staying engaged, knowing where the organization is going.

I think one of the great things about program management is, you are often aware of some of the other initiatives that are going on in the organization, because, many times, you're listening to them in different status meetings with higher levels in the organization. You kind of know where the winds are blowing and how you might be able to provide even more value in your program by knowing that.

Samir: What a great answer to what does it mean to be a good program manager? Thank you so much. Thank you for your time and thank you for sharing with us your expertise and your knowledge on program management.

Brian: Yes. Samir, thank you very much. I enjoyed it very much and best wishes to you on your book.

Appendix 3

AGILE PRIMER

Agile development is growing in popularity as it results in faster time to market, and in most cases better software. Agile teams build software incrementally in short iterations called *sprints*. Sprints are usually one to four weeks long. Instead of a single, long development cycle, the goal in agile projects is to build software incrementally. It fosters closer collaboration, better risk management, and results in a working product in a short time. The measure of progress for an agile project is working software at the end of a sprint.

Typical roles on an agile project are Product Owner, Scrum Master, and the team. The Product Owner is like a project sponsor, who is responsible for the product backlog. The product backlog is like a big list of requirements. For each sprint, the team jointly selects the items from the product backlog to work on and build software during that sprint. The Scrum Master's primary job is to remove all roadblocks. This close collaboration ensures that the entire team is working on the most valuable requirements.

Planning on agile projects also follows an iterative planning cycle. Before every sprint, the team jointly does sprint planning for the next sprint. At the end of a sprint, the team demos its working product, conducts retrospective sessions to understand what works well, and what requires improvement. This cadence of backlog grooming, sprint planning, demo, retrospective, next sprint planning and release planning is repeated until the project is done.

Agile teams touch base daily for a quick Scrum daily meeting, and get in sync with everyone. The rapid feedback mechanism reduces risk on projects as well as facilitates transparent communication among the team members. If you are new to agile, you want to start by reading the agile manifesto. As agile becomes mainstream, it is a skill that you can no longer ignore.

FIGURE A3.1
Appendix_ScrumAgileVisual.

Appendix 4

BENEFITS MANAGEMENT*

Anfre Toso Arrivabene, MBA, PMP, PgMP

INTRODUCTION

Programs are considered to be finished after all deliverables are completed and formally accepted, and all formal closure procedures are performed. However, it is becoming clear to many organizations that the quality of the program and its associated project deliverables and the quality of the execution in terms of budget and schedule compliance and customer satisfaction is not by far any guarantee of project success. While project success is measured by "product and project quality, timeliness, budget compliance, and degree of customer satisfaction" (PMI, 2008, p. 11), the success of a program is measured by "the degree to which the program satisfies the needs and benefits for which it was undertaken" (p. 11). Many programs and projects lack the clear definition of what exactly is a benefit, how to measure its value, and, most importantly, how to assure its sustainment within the performing organization.

The purpose of this chapter is to clarify some aspects of the benefits management, establishing a relationship between the benefits life cycle and the program life cycle.

A benefit is an "outcome of actions and behaviors that provides utility to the organization" (PMI, 2008, p. 5), and, in most cases, contributes to at least one organizational goal, which is derived from the strategic map. The organizational strategy is a result of the strategic planning cycle and should be a reflection of the company's vision, mission, and cultural aspects that

---

* From Levin, G. 2012. *Handbook of program management: A life cycle approach.* Boca Raton, FL: CRC Press, Chap. 6. With permission.

are translated into a strategic plan or strategic map. The strategic plan then should be translated into an enterprise portfolio, which is a set of prioritized programs, projects, and other work, and is the link between the organization's strategy and the company's investments and initiatives. The ultimate goal of linking the portfolio with the organizational strategy is to establish a balanced and executable plan that will allow the organization to achieve its objectives. The management of the expected benefits begins with its proper identification and planning, provides an appropriate set of tools for tracking the realization of the benefits, and also assures that actions are taken in order to guarantee that the resources to sustain the benefits are in place after their realization.

There are many reasons why programs and projects fail, and the causes go beyond the traditional budget, schedule, and quality problems. Shenhar and Dvir (2007) stated that even if we adhere to a project plan, we may not achieve the project's long-term business goals. A program—and, in fact, any enterprise investment—should be ultimately evaluated by how successfully the intended benefits are delivered and effectively turned into business results. One of the main reasons projects and programs fail is the inability to deliver and/or sustain the expected business benefits that justify the investment of company resources in the program. Although the project management maturity level in organizations has showed a significant improvement in the last decade, recent research shows that only 15% of senior-level project and program managers keep track of programs and projects benefits realization (ESI International, 2011).

BENEFITS CATEGORIES

In order to facilitate and support its identification and further analysis, benefits can be classified into key categories. Some benefits are easily determined and quantified, for example, increases in revenues or cost reductions. There are some benefits, though, that are more difficult to define because their contribution to business results are not direct and easily measured. Williams and Parr (2004) defined three main categories of benefits

- Direct or Tangible Benefits
  - *Financial Benefits*: Benefits that can be measured into currency quantities, such as increases in revenue and cost reductions.

- *Nonfinancial Benefits*: Benefits that are measurable, but not in currency or financial terms, such as client retention and lower staff turnover.
- Indirect or Intangible Benefits
- Benefits that are not easily quantified and measured, such as customer satisfaction, corporate image, and access to new markets.

BENEFITS MANAGEMENT LIFE CYCLE

In order to provide a predictable and coordinated manner to manage the expected program benefits and to assure compliance with governance standards, it is useful to establish a set of "processes and measures for tracking and assessing benefits throughout the program life cycle" (PMI, 2008, p. 20). The delivery of benefits that in some level provide new capabilities or that improve existing ones is one of the main critical success factors of an investment, whether it is carried out as a project or as a program.

The identification, qualification, and further quantitative analysis of the desired business benefits—aligned to the business strategy—should precede and direct the decisions to proceed with the program efforts along the program life cycle and also establish the baseline for measuring the program progress and success.

There are four phases for the benefits management life cycle, running in parallel with the program life cycle. It is important to emphasize that at the very beginning of the benefits management life cycle there should be a link to the company's strategy plan in a way that it provides a clear identification of the desired business benefits and in order to allow the prioritization of the program initiative within the context of the enterprise portfolio. Determining how the proposed benefits are aligned to the business strategy and also how relevant those benefits are in the enterprise environment is an important tool to clarify the justification for investing in the program.

The benefits management life cycle proposed by the Project Management Institute (PMI) flows in a manner that the level of information and maturity regarding the benefits management process increases along with the development of the program life cycle, from the very early defining and planning stages throughout the benefits realization, transition to operations, and final program closure. The next sections will provide some clarification on the four stages of the benefits management life cycle.

Benefits Identification

At the early stages of the program life cycle, most of the efforts are directed to the identification and qualification of the expected business benefits. The program business case or the program mandate should clarify the external and internal forces that drive the organization to the need for a change, which may lead to a program being created. During the identification of the expected benefits, it is important to establish a "clear definition and agreement among stakeholders on the factors contributing to these identified benefits" (PMI, 2008, p. 20). Some factors are critical for appropriate benefits identification at this stage of the benefits management life cycle:

- Understanding the business strategy
- Mapping each of the expected benefits into the strategic plan and goals
- Quantifying the benefits (estimates)
- Determining the *core* expected benefits
- Comparing the current state to the expected postprogram scenario
- Comparing the current state to the projected scenario in case the program is not implemented, if appropriate
- Determining the premises for estimating the benefits

The benefits identification phase should strongly take into account the organization's strategic plan in order to establish a clear alignment with the company's long-term objectives. Ideally, each of the expected benefits should be mapped into the company's strategic plan. At this stage, there is probably not much data supporting the quantitative analysis for the benefits, so it is most likely that order-of-magnitude estimates are being used. Figure A4.1 presents an example of a benefit identification and strategic mapping for a CO_2 emissions reduction program.

Figure A4.2 is a graphical representation of the comparison between the program's contribution to the corporate Key Performance Indicators (KPI) and a projection of the expected scenario if the program is not undertaken.

It is important to emphasize the link between the benefits and the business strategy is dynamic and is affected by changes to either or both. Any changes in the company's strategy should serve as a trigger to review the program goals and expected benefits and that could potentially cause its

Strategic Map					Program			
							Cumulative Reductions/ Year	
Customer Outcome	Strategic Objective	Key Performance Indicator	Current Value	Goal for KPI	Estimate Program Contribution	Year 1	Year 2	Year 3
To be perceived as a more eco- friendly company	Reduce the CO2 activity emissions	Total CO2 emissions/ year	1.5 ton/ year	0.8 ton/ year	Reduction of 0.5 ton/year	0.1 ton	0.3 ton	0.5 ton

FIGURE A4.1
Example of benefits identification and mapping. (From Levin, G. 2012. *Handbook of program management: A life cycle approach.* Boca Raton, FL: CRC Press. With permission.)

early termination. On the other hand, changes to the program that affect its capacity to deliver the planned benefits within schedule, budget, and capacity should be monitored and escalated to the appropriate governance structure in order to assure the necessary alignment to the business strategy.

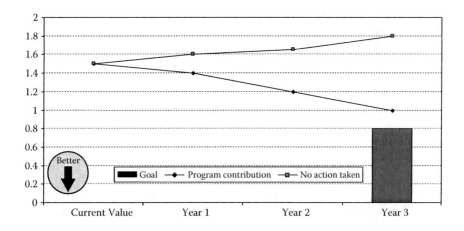

FIGURE A4.2
Example of benefits identification. (From Levin, G. 2012. *Handbook of program management: A life cycle approach.* Boca Raton, FL: CRC Press. With permission.)

Benefits Analysis and Planning

After the expected program benefits are identified and agreed upon, the groundwork for defining the tactical means by which the benefits are to be achieved takes place. The analysis of the program's expected benefits should result in a supportive basis for planning the benefits realization, providing information that will support further planning decisions. The benefits planning process is iterative and should be revisited at predefined check points, such as program phase-gate reviews; at points where there are scope changes; or when risks occur. Some relevant information to take into account during the benefits planning process includes:

- The prioritization of the expected program benefits and the estimated contribution from each component in achieving the program benefits. This information will support the planning decisions regarding the prioritization of the program components and the definition of the program scope.
- The benefits metrics and quantitative analysis that will support the monitoring of the benefits realization.
- The benefits measurement and completion criteria.
- The roadmap for delivering the expected benefits.

There are four main activities for the benefits analysis and planning phase (PMI, 2008):

1. *Derive and prioritize the program's components*—to identify each component's contribution to the expected program benefits and prioritize them accordingly.
2. *Derive benefits metrics*—in order to allow its further measurements during the program's execution and after its transition to ongoing operations.
3. *Establish a benefits realization plan and methods to monitor it*—to support program execution and the benefits realization tracking and reporting with an emphasis on resources needed, benefits interdependencies, realization premises, and constraints.
4. *Map benefits into the program plan*—to establish a correlation between the program's expected benefits and those of its constituent components.

Prioritizing Program Components

In order to derive and prioritize the program components, it is necessary to determine the contribution from each component to the overall expected program benefits. A *Component — Benefits* correlation matrix is proposed in Figure A4.3.

Once the contributions from each component are determined, it is possible to prioritize the components using a weighted decision matrix. Each benefit is associated with a weight, according to the program benefits prioritization criteria determined previously and agreed upon with the program sponsor and the main program stakeholders. For each individual component, it is possible then to calculate the overall contribution to the program objectives as follows:

$$Score \rightarrow \sum_{i=1}^{\#benefits} Contribution_i * Weight_i$$

For the example illustrated on Figure A4.3, the weights for each proposed program benefit were determined as follows as shown in Figure A4.4.

Taking, for example, the contributions from Project "A" to each benefit represented in Figure A4.4, and applying the formulae to the weights determined in Figure A4.5, we have the following results:

Applying the same calculations to projects B, C, and D on Figure A4.5, the result is a program weighted Component Benefit Matrix, represented in Figure A4.6.

	Expected Program Benefits		
Component	Reduction of CO2 Emissions	Cost Reduction	Operational Safety Improvement
Project A	0.5 MM ton/year	USD 1.5 MM/year	15%
Project B	2.5 MM ton/year	USD 0.4 MM/year	25%
Project C	1.5 MM ton/year	USD 3.5 MM/year	0%
Project D	1.5 MM ton/year	0	5%
Program	**6 MM ton/year**	**USD 5.4 MM/year**	**45%**

FIGURE A4.3
Component benefit analysis. (From Levin, G. 2012. *Handbook of program management: A life cycle approach.* Boca Raton, FL: CRC Press. With permission.)

Benefit	Weight
Reduction of CO2 emissions	5
Cost reduction	3
Operational safety improvement	1

FIGURE A4.4
Example of weighted benefit matrix. (From Levin, G. 2012. *Handbook of program management: A life cycle approach.* Boca Raton, FL: CRC Press. With permission.)

In the example above, based on the results of the weighted component benefit matrix, the prioritization of the program components should be (1) Project B, (2) Project A, (3) Project C, and (4) Project D. The component prioritization analysis will guide future decisions during the program execution and is especially important when it comes to decisions regarding the allocation of shared or conflicting resources.

Benefit	Benefit Weight	Project Contribution	Weight
Reduction of CO2 emissions	5	0.5	2.5
Cost reduction	3	1.5	4.5
Operational safety improvement	1	15	15
Total Project Score			**22**

FIGURE A4.5
Calculation of project score. (From Levin, G. 2012. *Handbook of program management: A life cycle approach.* Boca Raton, FL: CRC Press. With permission.)

	Program Benefits			
	Reduction of CO2 Emissions	Cost Reduction	Operational Safety Improvement	Component Score
Weight	5	3	1	
Project A	2.5	4.5	15	22
Project B	12.5	1.2	25	38.7
Project C	7.5	10.5	0	18
Project D	7.5	0	5	12.5

FIGURE A4.6
Weighted component benefit analysis. (From Levin, G. 2012. *Handbook of program management: A life cycle approach.* Boca Raton, FL: CRC Press. With permission.)

Note that the program components are prioritized according to their individual contributions to the overall program benefits. Other factors, however, should be taken into account when sequencing the components for program execution, such as resource availability, funding constraints, regulatory demands and components interdependencies, which are not considered in this example.

Developing a Benefits Realization Roadmap

In order to provide a baseline for monitoring program execution and to serve as a monitoring parameter, a roadmap for the benefits realization is required. Once the program components are prioritized, and a high-level program schedule is developed, it is possible to establish a time frame in which the benefits are going to be delivered and realized.

Based on each individual component contribution for the benefit "Reduction of CO_2 Emissions," from the example of the Component Benefit matrix represented in Figure A4.6, a time frame for the realization of the overall program benefits can be developed as follows and shown in Figure A4.7.

Figure A4.8 shows the graphic representation of the quarterly and cumulative benefits realization roadmap, based on the distribution represented in Figure A4.8.

A benefits realization roadmap as illustrated below is an important tool to keep track of the benefits delivery, and it also serves as an effective communication tool in order to clearly set the stakeholder's expectations for both the program and individual components. Program managers should

	Year 1				Year 2				Year 3				Total Component Contribution
	Q1	Q2	Q3	Q4	Q1	Q2	Q3	Q4	Q1	Q2	Q3	Q4	
Project A			0.3			0.1		0.1					0.5
Project B				0.5			1.5			0.2		0.3	2.5
Project C					0.3			0.6			0.6		1.5
Project D		0.3				0.8			0.4				1.5
Total	0	0.3	0.3	0.5	0.3	0.9	1.5	0.7	0.4	0.2	0.6	0.3	
Cumulative	0	0.3	0.6	1.1	1.4	2.3	3.8	4.5	4.9	5.1	5.7	6	

FIGURE A4.7
Program benefits roadmap. (From Levin, G. 2012. *Handbook of program management: A life cycle approach.* Boca Raton, FL: CRC Press. With permission.)

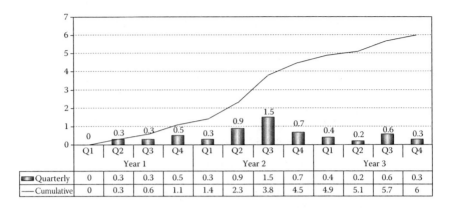

	Q1	Q2	Q3	Q4	Q1	Q2	Q3	Q4	Q1	Q2	Q3	Q4
		Year 1				Year 2				Year 3		
▣ Quarterly	0	0.3	0.3	0.5	0.3	0.9	1.5	0.7	0.4	0.2	0.6	0.3
— Cumulative	0	0.3	0.6	1.1	1.4	2.3	3.8	4.5	4.9	5.1	5.7	6

FIGURE A4.8

An example of a graphic representation for the benefits roadmap. (From Levin, G. 2012. *Handbook of program management: A life cycle approach.* Boca Raton, FL: CRC Press. With permission.)

assure agreement on the benefits roadmap with the program sponsor, customer, and other relevant stakeholders as appropriate. The roadmap should be part of the program and individual component's business cases, and project managers should have it as an input for setting up their project plans. Once agreed upon, the roadmap serves as a baseline for tracking and reporting the benefits realized during the program execution and should be used later during the life cycle to compare the actual performance to the plan.

Performing Benefits and Risk Analysis

Another important factor to be taken into consideration during the benefits analysis and planning phase is the benefits versus risk analysis, which is performed in order to assess the risks affecting the delivery of the program's benefits. One approach that is often overlooked is to establish a correlation between the program risks and the program benefits, in addition to mapping the risks to the program components (PMI, 2008). A correlation matrix that allows mapping and assessing the impact of identified program risks into each expected benefit is proposed in Figure A4.9 as follows:

The Benefit Risk correlation matrix can help the elaboration and prioritization of the program risk response plans because it assesses the impact of each identified risk to the realization of the expected program benefits. This assessment, combined with the program benefits prioritization

Program Risks	Expected Program Benefits		
	Reduction of CO2 Emissions	Cost Reduction	Operational Safety Improvement
Risk 1	High	—	—
Risk 2	Medium	High	—
Risk 3	—	Low	High
Risk 4	Low	High	—

FIGURE A4.9

Benefit × risk correlation matrix. (From Levin, G. 2012. *Handbook of program management: A life cycle approach.* Boca Raton, FL: CRC Press. With permission.)

criteria, is a powerful tool to guide the allocation of contingency reserves to program components.

Performing Benefits/Cost Analysis

Whether a benefit is measurable in monetary terms or not, an assessment of the benefits versus the cost of each of the program's components is useful in order to help the program manager prioritize the components in the program and also in order to establish a parameter for performance measurements in the later program life-cycle phases.

The most easy and traditional way to measure the benefit versus cost ratio is to simply divide the component cost (or the overall program cost) by the expected benefit, as illustrated in Figure A4.10.

In the example, what is being calculated is the ratio between each component cost and its contribution to the overall program benefit. Note that,

Component	Expected Program Benefits		
	Component Cost	Reduction of CO2 Emissions	Cost/Benefit Ratio
Project A	USD 2.0 MM	0.5 MM ton/year	USD 4.0 MM/(ton/year)
Project B	USD 8.0 MM	2.5 MM ton/year	USD 3.2 MM/(ton/year)
Project C	USD 6.0 MM	1.5 MM ton/year	USD 4.0 MM/(ton/year)
Project D	USD 9.0 MM	1.5 MM ton/year	USD 6.0 MM/(ton/year)
Program	**USD 25.0 MM**	**6.0 MM ton/year**	**USD 4.2 MM/(ton/year)**

FIGURE A4.10

Benefit/cost ratio calculation example. (From Levin, G. 2012. *Handbook of program management: A life cycle approach.* Boca Raton, FL: CRC Press. With permission.)

for example, the component "Project B" has a benefit/cost ratio—also called *specific cost*—of USD 3.2 MM/(ton/year), which means that, for "Project B," the reduction of one ton of CO_2 per year costs USD 3.2 MM. The same criteria may be applied to any quantifiable benefit.

The benefit/cost ratio also should be considered when determining the cost variance tolerances for the program and its components. For example, consider the component "Project D" in the example above. The component has a calculated benefit/cost ratio of USD 6.0 MM/(ton/year), which is the worst ratio of the project in comparison with the other three components, being 50% higher than the second worst ratio. Although a cost variance of, for example, plus 10% might be acceptable for the program as a whole, when the individual component's benefit/cost ratio is taken into consideration, program governance might consider it to not be an important variable. This information, however, may be used to select program component candidates as well as to decide whether to approve component initiation or to terminate an existing component.

Benefits Realization Monitoring Plan

In order to establish a framework for agreement on the benefits realization measurement criteria, the program manager should establish a plan for benefits measurement after the program implementation. The plan should refer to any preexisting KPIs for future reference, if one already exists, and clearly determine who is responsible for the benefits measurements and communication. Some benefits might be difficult and expensive to measure, so any resources needed to conduct the benefits measurements should be determined, such as human resources, new equipment, or facilities. The plan should be formally registered as a deliverable at the program level and communicated and agreed upon by the main stakeholders.

For each benefit, the plan should determine the criteria and requirements for measuring the benefits realized and should contain information, such as:

- Benefit identification
- Measurement units: In order to allow a clear understanding of the measurements and to allow the further comparisons to the current KPI values, such as monetary units, meters, tons, etc. The measurement unit for the benefit should be the same as the KPI to which it will be compared.

- Date (or event) to start the benefit measurements: It is important to determine when the measurements are going to start, whether at a predetermined date, or after a specific program event, such as a new equipment ramp-up period or a regulatory demand.
- Frequency of measurements (monthly, quarterly, etc.)
- Measurement criteria and methods
- Resources needed for measurements: Any resources needed in order to perform the measurements should be estimated and planned. Some benefits might demand extra resources to be tracked and measured, and it may be necessary to acquire measurement equipment and facilities or contract for specialized services in order to measure the benefits.
- KPI previous value (if one exists): Any KPI affected by the program implementation and related to the benefit being measured should be tracked before the program implementation for future comparisons.
- Expected KPI value: The new value expected for the KPI after the delivery of the benefit.
- Communication methods: How the benefits measurements are going to be communicated to the stakeholders— a monthly report, a meeting, etc.
- Measurement responsibility

Benefits Realization

In addition to monitoring the performance of the program and its constituent components in terms of budget, schedule, and quality, an effective governance structure should be in place to assure periodic monitoring of the expected benefits delivery. In some cases, benefits will be delivered incrementally during the program execution, while others will be delivered only after full implementation of the program's scope. In both cases, a set of procedures and methods for monitoring and reporting benefits should be in place. Also, the program's business case should be revisited at the periodic performance reviews. Using the business case as a constant reference during the benefits monitoring activities allows the program manager to periodically check the program alignment with the original expected benefits.

Depending on the nature of the program, the delivery of its intended benefits could fall into one of the following categories (Williams and Parr, 2004):

- Sudden—Program benefits are delivered at once or after a short ramp-up or stabilization period.
- Incremental—Program benefits start to be delivered before program conclusion, and each program component's contribution adds to the overall program benefits being delivered.
- Transient—Benefits have a time frame during which they can be realized. After a certain time period, the benefits realization starts to fade or is minimized. This could be related to a commercial window of opportunity, a regulatory demand, or a technology issue.

There are three main activities for the Benefits Realization phase of the benefits management life cycle (PMI, 2008):

Monitoring Components

Monitoring the execution of the program components provides program managers with the necessary information to forecast the future realization of the expected program benefits. Special attention should be devoted to program changes that could affect the benefits realization or viability. Also, deviations on components schedules or budgets might cause delays in realizing benefits, which might even cause the program to be canceled.

Figure A4.11 illustrates a deviation in a benefit realization forecast because of delays in a component schedule. Figure A4.12 illustrates a deviation in a benefit realization forecast because of a component scope or capability reduction. As an example, Figure A4.13 illustrates a program's benefit realization forecast against the benefits realization plan baseline. The updated forecast for the contribution from each program component is compared against the expected contribution stated on the program and project business cases. Negative deviations on the program benefits forecasts should serve as a trigger to perform a root cause analysis and subsequent corrective actions in order to bring the benefits realization back to the original roadmap. These corrective actions should be implemented by the components' project managers and followed up by the program management team and the program manager.

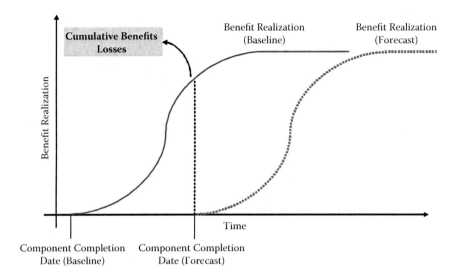

FIGURE A4.11
Benefit forecast deviation because of a component schedule delay. (From Levin, G. 2012. *Handbook of program management: A life cycle approach.* Boca Raton, FL: CRC Press. With permission.)

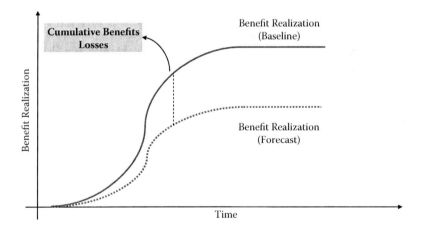

FIGURE A4.12
Benefit deviation because of a component scope or capability reduction. (From Levin, G. 2012. *Handbook of program management: A life cycle approach.* Boca Raton, FL: CRC Press. With permission.)

| | Expected Program Benefits | | | | |
Component	Planned	Forecast	Deviation	Cause Analysis	Corrective Actions
Project A	0.3 MM ton	0.5 MM ton	0.2 MM ton		
Project B	1.5 MM ton	0.0 MM ton	(1.5 MM ton)		
Project C	0.5 MM ton	0.4 MM ton	(0.1 MM ton)		
Project D	1.0 MM ton	1.5 MM ton	0.5 MM ton		
Program	3.3 MM ton	2.4 MM ton	(0.9 MM ton)		

FIGURE A4.13

Forecasting program benefits. (From Levin, G. 2012. *Handbook of program management: A life cycle approach*. Boca Raton, FL: CRC Press. With permission.)

Maintaining a Benefits Register

Data collection regarding benefits realization should be performed periodically, in order to provide the program sponsor and stakeholders with the actual benefits realized versus the planned benefits. For those programs whose benefits are delivered before the formal program work is completed, the measurements for benefit realization should be performed in conjunction with the program periodic health checks and program performance measurements. The early program benefits realization measurement will support future governance decisions regarding the program's viability. There are even cases in which the program depends on those early expected benefits to help fund other program components or to adjust the program's scope and budget according to the actual realized benefits.

Keeping track of the actual program benefits achieved should take place during program execution and/or after program completion, depending on the nature of the program. Program managers should maintain a register of program benefits actually delivered by the program in order to evaluate the benefit's delivery performance against the benefits realization plan (PMI, 2008). In that sense, a standardized set of benefits metrics should be established earlier in the benefits management life cycle in order to prepare a coherent and cohesive program benefits register. In order to allow future comparisons and analysis between planned and actual benefits, the metrics for measuring and registering realized benefits should be the same metrics used to set the benefits baseline and also the same as those used to define the goals for corporate KPIs. This approach will allow the program team to communicate the program results in terms of their contribution to the overall organization's goals.

Program Name:
    Component: Component A
       Benefit: Reduction of CO_2 emissions					

Date of Measurement	Planned	Actual	Deviation	Cause Analysis	Corrective Actions

FIGURE A4.14

Component benefit register. (From Levin, G. 2012. *Handbook of program management: A life cycle approach.* Boca Raton, FL: CRC Press. With permission.)

Figure A4.14 illustrates an example of a register from a program component's benefit measurement. Note that the measurements are periodical, and for each measurement, there is a comparison with the planned benefit value expected for that particular period and an analysis of the deviations, with their respective cause analysis and corrective actions.

The benefit deviation analysis could be complemented with an assessment of the cumulative values expected to date and a comparison with the global expected KPI related to the benefit, as illustrated in Figure A4.15.

Program managers also should, within the context of the program governance structure, assess the benefits realized in relation to the overall program performance. The aspects involved in the benefits realization assessment should include, but not be limited to (PMI, 2008):

- Program resources spent to date as compared to the actual benefits achieved.
- Updates on the resources needed in order to deliver the remaining expected benefits.

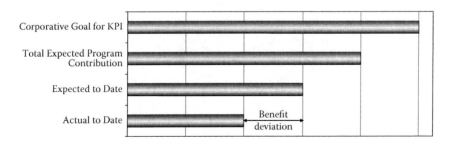

FIGURE A4.15

Benefit measurement assessment example. (From Levin, G. 2012. *Handbook of program management: A life cycle approach.* Boca Raton, FL: CRC Press. With permission.)

- Governance decisions that impact the benefits realization, whether negative or positive.
- Strategic alignment of the expected benefits to the organization's goals and objectives.
- Updates on the status of any risks associated with realizing the benefits.
- The overall program and its constituent component's performance, in terms of budget, schedule, quality, and resource allocation.

When assessing the benefits realization performance, one piece of information that is often overlooked is the actual benefit versus cost ratio for the program benefits. The traditional approach is to monitor the cost performance for the program components and its deliverables, so there is a budget control that operates within a predetermined and approved tolerance range. But, what if there is a cost variance that falls within the tolerance range, but affects the benefit versus cost ratio in such a way that makes the program or one of its components no longer viable? Or what if there is not a cost overrun, but the actual benefits delivered by the program are below the expected benefits, affecting in the same fashion the benefit versus cost ratio?

In order to allow a more effective and benefit-driven approach to program cost control, the assessment of the actual benefit versus cost ratio for each of the program benefits and components should be performed during the delivery of the program benefits phase. Governance decisions regarding component initiation approvals, program scope changes, and program go/no-go decisions should consider the benefit versus cost analysis.

For the example illustrated on Figure A4.16, although there are some significant cost variances for each component, the actual benefits achieved by the components either kept the benefit versus cost ratio within the planned ranges or made an improvement on the component's ratio—except for the component "Project C." When the overall program ratio is calculated, the benefit versus cost ratio in terms of USD MM per ton/year is better than originally planned.

Benefits Reporting

Program benefits may be realized before the program work is completed (PMI, 2008). In order to keep the sponsor and stakeholders' expectations about the program benefits within a realistic range, program managers

	Expected Program Benefits				
Component	**Planned Cost/ Benefit Ratio**	**Component Actual Cost**	**Actual Reduction of CO2 Emissions**	**Actual Cost/Benefit Ratio**	
Project A	USD 4.0 MM/ (ton/year)	USD 2.5 MM	0.5 MM ton/year	USD 5.0 MM/ (ton/year)	Worse
Project B	USD 3.2 MM/ (ton/year)	USD 7.0 MM	3.5 MM ton/year	USD 2.0 MM/ (ton/year)	Better
Project C	USD 4.0 MM/ (ton/year)	USD 8.0 MM	1.3 MM ton/year	USD 6.2 MM/ (ton/year)	Worse
Project D	USD 6.0 MM/ (ton/year)	USD 11.0 MM	1.8 MM ton/year	USD 6.0 MM/ (ton/year)	Same
Program	**USD 5.4 MM/ (ton/year)**	**USD 28.5 MM**	**7.1 MM ton/year**	**USD 4.0 MM/ (ton/year)**	Better

FIGURE A4.16
Actual benefit versus cost assessment example. (From Levin, G. 2012. *Handbook of program management: A life cycle approach*. Boca Raton, FL: CRC Press. With permission.)

should periodically assess and report the actual realized program benefits. Among other factors, benefits reporting should take into account:

Program Business Case
Benefits Realization Plan
Benefits Roadmap
Actual Benefits to Date
Program Stakeholder Analysis
Program Communications Management Plan
Communications Strategy
Program Communications Log
Program Master Schedule

Benefits Transition

The final step in the life cycle is to formally transition the benefits to the program's customers, end users, or a product or support group within the organization. At this time, the program has closed, and its benefits have been realized successfully according to its stakeholders, the governance board, and the program sponsor. The program's transition plan has been executed, and stakeholders who now are responsible for sustaining the benefits have been active participants in the transition process.

SUMMARY

This chapter has described a key aspect of program management, if not the most important part of program management—benefit realization. It has presented through the use of examples of the components of the program life cycle so the program manager has guidelines to follow as he or she plans, executes, monitors, and closes his or her program to ensure the program's stated benefits are realized and hopefully exceeded.

REFERENCES

ESI International. 2011. *The global state of the PMO: Its value, effectiveness and rule as the hub of training*. Arlington, VA: ESI International.

Project Management Institute (PMI). 2008. *The standard for program management*, 2nd ed. Newtown Square, PA: Project Management Institute.

Shenhar, A. J. and D. Dvir. 2007. *Reinventing project management: The diamond approach to successful growth and innovation*. Boston, MA: Harvard Business School Press.

Williams, D. and T. Parr. 2004. *Enterprise programme management: Delivering value*. Hampshire, UK: Palgrave Macmillan.

Appendix 5

PROGRAM MANAGEMENT OFFICE: ROLE AND CHALLENGES*

Monica Gaita, MBA, PMP, PgMP

INTRODUCTION

A program, defined as a subset of the company's portfolio, comprising projects, operational work, and sometimes other programs, has many of the features and faces some of the complexities of the entire organization.

Program management derives its value by managing in an integrated and synergistic way multiple-related projects. The entity, which sets the common language, makes the links, and provides basis for management decisions, is the Program Management Office (PMO).

As the *Standard for Program Management*, 2nd edition (PMI, 2008) states, the core of the PMO activity is

- "defining the program management processes that will be followed;
- managing schedule and budget at the program level;
- defining the quality standards for the program and the program's components;
- providing document configuration management; and
- providing centralized support for managing changes and tracking risks and issues" (p. 11).

Additionally, the PMO can create value for the organization by being actively involved in resource management, contract and procurement management, and other functions shared by the program's components.

---

* From Levin, G. 2012. *Handbook of program management: A life cycle approach*. Boca Raton, FL: CRC Press, Chap. 15. With permission.

This chapter will analyze the different PMO functions and show how each of them contribute to the performance of the program.

DEFINING THE PROGRAM MANAGEMENT PROCESSES

When part of a program, projects are not discrete, disjunctive initiatives, but are linked together through their role in delivering the program benefits. The PMO has a central place in monitoring benefits realization both through projects and other program components, such as operations.

Using an analogy to communication networks, the PMO might be seen as a switch enabling exchanging information and sharing resources among the components. In order for components to be understood by the network, they must use the same communication protocol. The communication protocol in this case is the program management methodology, represented by processes and materialized in specific tools and templates. Communication protocols use REQ (request) and ACK (acknowledged) flags within their messages. The PMO is the entity receiving and acknowledging requests originating in programs and projects. The switch could be configured to assign different priorities to requests, based on a set of predefined criteria and rules—just as the PMO does. Last, but not least, the switch produces standard and custom-made reports, which allow the administrator to optimize its configuration and the use of network resources.

Once it is established, the PMO's first priority is to develop and implement the standard methodology. The level of complexity of the processes must be scaled to the maturity of the organization. There is a fine line between too little and too much process. Continuing with the network communications analogy, the defined processes must be looked at as best routes from the program's start to the intended result. They are meant to avoid network clashes, loops, or missing data. In the same manner, the program management process is the highway leading to the final program objective.

However, the methodology must be dynamic and driven by continuous improvement. A too rigid set of rules for managing the programs and their related projects could lead to inefficiencies and even blockages. This is the case where the processes are not fulfilling their support role, but are acting as strong constraints instead. The focus has moved from the destination to the road itself. Program managers most probably do not feel valued and empowered to make the right decisions in their programs.

The amount of bureaucracy increases considerably with no corresponding created value. The overhead costs push the program's performance down. There is a considerable risk that the program manager and governance structures lose track of strategic goals while monitoring and controlling the nuts and bolts affecting the program and its associated projects' time, budget, and scope.

In the opposite extreme, a much too relaxed set of processes would not allow for proper program management. Opportunities for new projects under the program must be weighed against the same sets of criteria in order to allow for the selection of those initiatives that create more value to the organization. Project planning and resource allocation must be done in a coordinated way. As they are related, the projects within the program often share the same pool of human resources. The implementation plans must be synchronized as to allow for optimum utilization of people and skills. Risk management is the area where the impact of a missing defined process is most visible. Lessons learned are not shared within the program and with the rest of the organization. The cost of not having an effective and efficient PMO and standard methodology in place cumulates: non-conformance costs related to the products created and the management of the individual projects, higher costs of resources, and increased opportunity costs for unrealized efficiencies in the management of contracts and procurement are typical results.

Finding the right balance between rules and flexibility in the standard methodology is the goal of an effective PMO. The PMO should not only establish a methodology but also demonstrate flexibility in encouraging program and project managers to use it. The members of the PMO should be seen as critical drivers of business improvement—they must not just propose, implement, and apply a program management framework, but enable the company to improve its methodologies and strategic management maturity. The PMO staff must implement the program strategy by creating policies and procedures, acting as a single point of contact for initiation and support of projects throughout their execution, and assisting project managers when needed. The PMO must continuously watch for gaps in project planning, delivery, and risk assessment processes that lead to problems and nonconformance costs and address them with improvements of the methodology.

SELECTION OF PROGRAM COMPONENTS

The PMO defines and drives the selection of program components. Project selection must follow a standard methodology to ensure effective management of resources and alignment with strategic goals.

Criteria must be approved and the process fully supported by the leadership team. While return on investment (ROI) has its permanent place on this list, there are others that reflect the type of organization and its strategic direction. When developing the balanced scorecard for project selection by the program governance committee, the PMO also should consider the:

- weight of strategic alignment criteria versus that of the ROI;
- intangible benefits, such as improved working conditions, ease of use, workforce satisfaction, and improvement in corporate morale;
- dependencies and impact on other programs and projects in the organization;
- synergistic opportunities;
- business continuity aspects; and
- legislation and regulatory standards.

This process must be tailored to the company, and the PMO must routinely monitor active and proposed projects, and, when appropriate, their programs, against corporate strategy.

MANAGING SCHEDULE AND BUDGET
AT THE PROGRAM LEVEL

The PMO supports the program manager in developing the program schedule and plays an active role, therefore, in monitoring and controlling it. At the program level, the planning will remain at a high level with only the component milestones, which represent an output to the program or interdependencies between projects being shown and tracked.

The PMO does not play a central role but participates in planning the program. One of its key roles is to facilitate the efficient use of project resources, but also operational resources. In order for the program schedule to be feasible and optimum, it must account for availability of skilled resources. When approved business cases for component projects define headcount requirements for further operating and supporting the product created after its go live date, the PMO must ensure those roles are staffed. If they cannot be staffed appropriately with people with the key competencies, knowledge, and skills, it becomes a widespread issue leading to inefficiencies within programs.

When planning the initiation of components within the program, the PMO must additionally consider dependencies on other projects, synergistic opportunities, and contribution to the program's bottom line; the ROI; availability of program funding; and benefits analysis. The components' detailed plans will then support the higher-level program master plan.

The PMO will ensure that progress reports of components follow a defined template, permitting easy aggregation and compilation of data at the program level. Most frequently, PMOs require measurable status on milestones, percent complete, and actual versus planned efforts and costs. As opposed to a portfolio status report, a program status report will show not only progress of components but also cross-impacts and interdependencies. Deviations in the delivery of one component will have a domino effect on other related components. Table A5.1 is proposed to monitor the program's performance.

It is the PMO's responsibility to propose to the project managers and program managers a governance structure than can effectively provide a clear process and suitable tools for managing the schedule and budget. Program metrics are defined, and earned value management is recommended to track progress.

DEFINING THE QUALITY STANDARDS FOR THE PROGRAM AND THE PROGRAM'S COMPONENTS

The PMO will ensure that the organization's quality management policy, as well as the overall intentions and direction of an organization with regard to quality, as formally expressed by top management, is materialized into the program quality management system. The program's quality system represents the organizational structure, responsibilities, procedures, processes, and resources needed to implement quality management.

TABLE A5.1

Monitoring Program Performance

Component #		Component Name		Component Manager		Approved Budget	
Gate 1		**Gate 2**		**Gate 3**		**Gate 4**		**Gate 5**	
Planned	Actual	Planned	Actual	Planned	Actual	Planned	Actual	Planned	Actual

Escalations, Risks, Issues, Decisions Pending	Impacted by Component #	Impacts on Component #	# of Change Requests	# of Incoming Claims	Customer Satisfaction	Contributes to Benefit #	SPI	CPI

Source: Levin, G. 2012. *Handbook of program management: A life cycle approach.* Boca Reton, FL: CRC Press. With permission.

The program quality management plan usually defines

- quality metrics for the product;
- quality metrics for processes; and
- what, when, and by whom is to be checked.

The nature of the quality assurance metrics that the PMO introduces into the program management environment heavily depends on the specifics of the program.

In some organizations, the quality manager function sits within the PMO; in others there is a standalone position in charge of quality planning, assurance, and control. In both cases, though, there is a close collaboration between the two. Each deviation from the quality standards established generates either rework or decreased customer satisfaction—both leading to postponed or unachieved final acceptance, hence, unrealized benefits.

Quality assurance metrics should include items such as quality variance values at interim checkpoints, number of project scope or solution changes, nonconformance costs associated with particular project teams and individual project managers, average customer acceptance rates and timing, and contribution of lessons learned to the program management knowledge base.

The quality manager must also verify the project managers' deliverables conform to the established program management methodology. The number of nonconformities found in component projects and the average timing of their resolution represent relevant quality metrics for the PMO.

Quality assurance is done for each important deliverable, but also at predefined stages in the projects. The gating process implies verifying the quality of the product's predefined milestones in the project and program life cycles. Gates have a common structure and consist of three main elements:

1. Inputs: Deliverables as per a checklist built for each gate
2. Criteria: Questions or metrics on which the prioritization and the stage gate decision is to be based (go/kill/hold/recycle)
3. Outputs: Results of the gate review—a decision (go/kill/hold/recycle), along with an approved action plan for the next gate and a list of deliverables and date for the next gate

FIGURE A5.1
Plan, do, check, act cycle applied to program management. (From Levin, G. 2012. *Handbook of program management: A life cycle approach.* Boca Raton, FL: CRC Press. With permission.)

The PMO has the responsibility to check that the documentation submitted for quality reviews is complete and is in the required format. In addition to information on component deliverables, information on component interfaces and interdependencies on other program components must also be included.

Quality assurance feeds back into program management processes topics of improvement, as an application of W. Edward Deming's (2000) "Plan–Do–Check–Act" continuous improvement framework, as shown in Figure A5.1:

> PLAN: Plan for improvements of the program management methodology
> DO: Implement the planned improvements on a small scale
> CHECK: Check to verify if changes produce the expected result
> ACT: Act to obtain the greatest benefits from changes

ROLE OF THE PMO IN RESOURCE MANAGEMENT

The PMO ensures that the program objectives are disseminated in the objectives of each project and that project managers' individual objectives are fully aligned. It plays an important role in evaluating the project managers' performance and assists in determining their career path within the organization. In order to ensure the program's overall performance, the PMO should base the project managers' appraisal process not only on budget, time, and scope, but also on realizing the benefits planned for his or her project as well as for the entire program.

However, aligning the project managers' objectives does not prove to be sufficient for granting positive results. The program governance board must work with the organization's top management to also align departmental objectives. By steering this alignment, the PMO can set the basis for collaboration toward meeting program's objectives versus working in silos to meet individual goals.

With programs spanning longer periods of time than projects, many components usually handled by the resource managers are taken over by the PMO under the direct supervision of the program manager. While other resources may be assigned on part-time or temporary basis, the key technical resources are often assigned to programs on full-time basis. With the full-time approach, the PMO shares with the department managers the responsibility of maximizing resource utilization across the program's components.

Staff requirements are first identified at program inception and, therefore, are revisited with each reporting cycle. The PMO will receive performance status reports, resource requirements, and resource releases from each of the components and will perform a five-step process described below and shown in Figure A5.2:

1. Aggregate resource requirements.
2. Check availability of resources within the program.
3. Negotiate resource allocations within the program, taking into account the performance of each component and its contribution to the realization of the program benefits.
4. Identify those resource needs that cannot be covered from the pool managed by the program and attempt to source them from the larger organization.
5. Perform resource allocation for the program components.

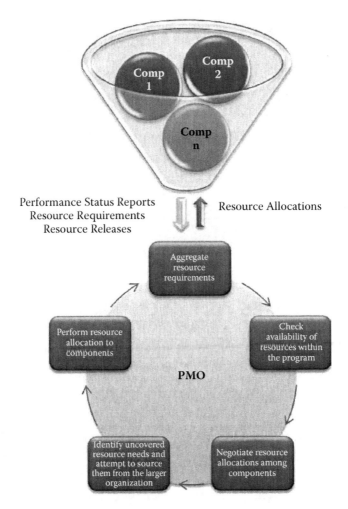

FIGURE A5.2
Resource allocation. (From Levin, G. 2012. *Handbook of program management: A life cycle approach*. Boca Raton, FL: CRC Press. With permission.)

THE PMO AND RISK MANAGEMENT

Risk management is one of the areas where managing related projects as a program as opposed to managing them individually creates tangible efficiencies.

Risk management planning at the program level includes strategies, tools, methods, reviews and reassessment processes, metrics gathering, standard assessment parameters, and reporting requirements to be used by each project in the program.

The component projects' risk registers are consolidated into a program risk register, and further analysis is done to determine the cross-impact of risks on the components and the overall program. Risks also are checked for redundancy, as the same possible event could be identified in one or more program components independently. Provisioning risk budgets within each of the components could artificially increase the perception of the overall program risk and could lead to biased management decisions. On the other hand, disregarding the direct or indirect effect of one risk event on other program components could lead to underevaluating the true risk impact.

The consolidated view might change the risks priority and justify changes in the response plans. Individual risk values are summed if they are encountered in multiple projects. The cost of preventive or contingent measures required to address the risks are optimized; therefore, the risk exposure calculated for the program is less than the sum of the components' risk exposures.

For example, consider a program with a goal to implement an integrated management system in 42 hospitals. Each hospital implementation is handled as a distinct project. The development of the core application and its centralized deployment also are managed as one project. The project manager for Hospital 1 identifies the following risk: "The emergency room (ER) medical staff might reject the new system, as they consider the user interface unfriendly." The proposed measure is to organize an additional training session for the ER staff in Hospital 1 and to offer onsite assistance in the ER for the first month after going live. Once the risk is raised to program level, it is recognized by all the other hospital projects. The cost of organizing additional training sessions and dispatching project resources onsite for 24/7 coverage in all 16 hospitals is calculated to be higher than the cost of improving the user interface by changing its layout and adding contextual help.

Risks should be distinguished between events having a potential adverse impact and those having a potential positive impact or opportunity. In the example provided above, implementing the mitigation strategy would lead to a secondary positive impact at the program level; the simplified and self-explanatory interface could be implemented on mobile devices as well.

PMO IN MONITORING AND CONTROL

PMOs have different roles in different companies, but most stakeholders perceive monitoring and controlling of project and program performance as the most important role of the PMO. The PMO relieves the management structures of the cumbersome task of consolidating performance data from a large number of projects. The PMO structures and compiles the information to offer the foundation for management decisions. Therefore, it is of utmost importance that the PMO strives to continuously assess and improve the relevance of its reporting to the executive levels of the program and the organization.

As part of the standard methodology, the PMO establishes the content, format, and frequency for progress reports from individual components within the program. The information gathered is then consolidated and presented to the program governance structure for analysis and decision making.

Program performance is tracked through a program scorecard in terms of scope, master schedule, program budget, stakeholder expectations, risks, procurements, issues, and also benefits metrics. The reports must offer more than traditional information on project progress and provide high-level information on each and every project. They also should show a roadmap about how each project addresses strategic objectives.

This consistent approach to monitoring and controlling across all projects within the program allows for timely identification of issues, efficient handling of escalations, and intervening in projects where needed. Additionally, it enables the program management to assess the level of stakeholder satisfaction. Monitoring and revalidating component projects for strategic alignment should be done as an iterative process, especially since under the current volatility of the global economy, goals targeted tend to change more frequently.

The template in Figure A5.3 can be used to monitor mapping of program components to planned benefits.

In some organizations, the PMO is mostly perceived as an administrative support function, while in other organizations, the PMO employs senior project and program managers who act as a first escalation level for project managers and support them in dealing with difficult situations in their projects.

Program Benefit #	Benefit Description	Measurement	Target	Program Component	Budget	Status [Proposed, Approved, In Progress, Completed, Rejected]

FIGURE A5.3
Template to monitor mapping of program components to planned benefits. (From Levin, G. 2012. *Handbook of program management: A life cycle approach.* Boca Raton, FL: CRC Press. With permission.)

ORGANIZATIONAL PERFORMANCE AND PMO PERFORMANCE

As stakeholders are heterogeneous, representing different departments, roles, and have a wide range of interests, their expectations from the PMO in the organizational context also are diverse.

The PMO's contribution to organizational performance is mostly viewed through its role of supporting program management. The PMO is supposed to be in control, setting rules, and proving its authority, but is expected to show flexibility and adaptability as well. When projects run smoothly and as planned, the PMO meetings are perceived as inefficient and a waste of time, leading to frustrated project managers who must redundantly provide project status information through multiple venues.

On the other hand, when projects encounter difficulties, the PMO must be able to perform critical analysis and be seen as problem solvers. The PMO staff must be adaptable and able to "hit the ground running" on any project where their involvement is required. Because of its integrative role in the organization, the PMO has a high-level view of the work under way and planned in the organization, and its staff members then act as a bridge between project managers, the program manager, and the executives. It also has a social role: it encourages communication, knowledge, best practices, and lessons learned sharing across the organization. Often staff members in a PMO must use negotiation skills in order to resolve conflicts escalated to their level, mostly related to the use of shared resources or opposing stakeholder interests.

When the strategic and important projects of the organization are completed with success, the contribution of the PMO is recognized. As Aubry and Hobbs (2011) found, justification of PMOs is often challenged: "… organizational performance is a subjective construct. … The organizational performance of PMOs will vary depending on who the evaluator is" (p. 4). For this particular reason, beside regular reports on the status of the program and its components, the PMO also must report regularly on its own activity and results to communicate and demonstrate its contribution and impact to the business.

It is critical, therefore, for the PMO leader to understand what the executive team values and how it can be delivered through the PMO. As Rad and Levin (2002) state: "… different types of PMOs solve different types of problems. Therefore, determining organizational objectives that are to be pursued as part of the PMO implementation and functions to be performed by the PMO is the first step in planning the implementation" (p. 157).

The PMO needs a clear charter and strong sponsorship within the organization, especially in weak matrices where project managers report to the different lines of business. The PMO's place in the organization must be clear with their responsibilities and limits of authority well defined and communicated by the program sponsor. The PMO must be empowered by program management to act on its behalf in order to achieve the goals established. The PMO charter is the document that formalizes this empowerment and establishes the PMO objectives.

In order to derive the PMO's objectives and measures for success, a stakeholder analysis must be performed with the participation of the:

Program governance board
Portfolio managers
Business unit managers
Functional managers
Project managers
Project controllers, etc.

The PMO's contribution to the program's performance could materialize in amount of benefits realized, improvements in gross margin compared to internal benchmarks, reduction of time in the project's life cycle, more effective resource utilization rates, unused risk contingencies, reduction in non-conformance costs, reduction in audit non-conformities, increased client satisfaction levels, more streamlined processes and less bureaucracy, etc. Not having a PMO in place would result in inconsistent reporting, errors that are due to incorrect project data, and unaddressed program risks.

SUMMARY

Created upon program initiation, the PMO is given through its charter a clear mandate, role, responsibilities, and limits of authority within the organization. While being in charge with setting the rules and ensuring compliance, the PMO must prove flexibility and be perceived as a supporting function rather than a constraining function.

The PMO staff performs administrative tasks in gathering status data from component managers and consolidating them, but it also must be ready to resolve escalations or step forward when difficulties in projects require it to do so. The PMO must be able to micromanage, but also to draw the big picture for upper management perusal. It creates links between component managers, with program management and stakeholders, and with functional managers across the organization and outside the organization when partners or subcontractors are involved.

It acknowledges and resolves requests for shared resources. It negotiates. If the program is a living structure, then the PMO is its pumping heart.

REFERENCES

Aubry, M. and B. Hobbs. 2011. A fresh look at the contribution of project management to organizational performance. *Project Management Journal 42* (1): 3–16.

Deming, W. E. 2000. *Out of the crisis.* Cambridge, MA: MIT Press.

Project Management Institute. 2008. *The standard for program management,* 2nd ed. Newtown Square, PA: Project Management Institute.

Rad, P. F. and G. Levin. 2002. *The advanced project management office: A comprehensive look at function and implementation.* Boca Raton, FL: St. Lucie Press.

Selected Bibliography

Daily Infographic blog. Online at www.dailyinfographic.com

Eventual Millionaire, The. Online at: www.eventualmillionaire.com

Future of Project Management blog. Online at: www.futureofprojectmanagement.com

Levin, G. 2012. *Handbook of program management: A life cycle approach*. Boca Raton, FL: CRC Press.

Project Management Institute. 2013. *The standard for program management*, 3rd ed. Newtown Square, PA: PMI.

Index